The purpose of this book is to formulate a way of thinking about issues of power, moral identity, and ethical norms by developing a theory of responsibility from a specifically theological viewpoint; the author makes clear thereby the significance for Christian commitment of current reflection on moral responsibility. The concept of responsibility is relatively new in ethics, but the drastic extension of human power through various technological developments has lately thrown into question the way human beings conceive of themselves as morally accountable agents. It is this radical extension of power in our time which poses the need for a new paradigm of responsibility in ethics. Schweiker engages in an informed way with what is therefore a highly topical discussion. By developing a coherent theory of responsibility, and inquiring as to its source, the author demonstrates the unique contribution which might be made by Christian thought to moral questions in the next century.

RESPONSIBILITY AND CHRISTIAN ETHICS

NEW STUDIES IN CHRISTIAN ETHICS

General editor: Robin Gill

Editorial board: Stephen R. L. Clark, Anthony O. Dyson,
Stanley Hauerwas and Robin W. Lovin

In recent years the study of Christian ethics has become an integral
part of mainstream theological studies. The reasons for this are not hard
to detect. It has become a more widely held view that Christian ethics is
actually central to Christian theology as a whole. Theologians increasingly have had to ask what contemporary relevance their discipline has
in a context where religious belief is on the wane, and whether Christian
ethics (that is, an ethics based on the Gospel of Jesus Christ) has
anything to say in a multi-faceted and complex secular society. There is
now no shortage of books on most substantive moral issues, written
from a wide variety of theological positions. However, what is lacking
are books within Christian ethics which are taken at all seriously by
those engaged in the wider secular debate. Too few are methodologi-
cally substantial; too few have an informed knowledge of parallel
discussions in philosophy or the social sciences. This series attempts to
remedy the situation. The aims of New Studies in Christian Ethics will
therefore be twofold. First, to engage centrally with the secular moral
debate at the highest possible intellectual level; second, to demonstrate
that Christian ethics can make a distinctive contribution to this debate –
either in moral substance, or in terms of underlying moral justifications.
It is hoped that the series as a whole will make a substantial contribution
to the discipline.

A list of titles in the series is given at the end of the book.

RESPONSIBILITY AND CHRISTIAN ETHICS

WILLIAM SCHWEIKER

Associate Professor of Theological Ethics,
The University of Chicago

CAMBRIDGE
UNIVERSITY PRESS

CAMBRIDGE UNIVERSITY PRESS
Cambridge, New York, Melbourne, Madrid, Cape Town, Singapore,
São Paulo, Delhi, Dubai, Tokyo

Cambridge University Press
The Edinburgh Building, Cambridge CB2 8RU, UK

Published in the United States of America by Cambridge University Press, New York

www.cambridge.org
Information on this title: www.cambridge.org/9780521657099

First published 1995
Reprinted 1997
First paperback edition 1999

A catalogue record for this publication is available from the British Library

Library of Congress Cataloguing in Publication data
Schweiker, William.
Responsibility and Christian ethics / William Schweiker.
p.   cm. – (New studies in Christian ethics)
Includes bibliographical references.
ISBN 0 521 47527 9 (hardback)
1. Responsibility.   2. Christian ethics. I. Title.   II. Series.
BJ1451.S344   1995
241–dc20 95–45193   CIP

ISBN 978-0-521-47527-3 Hardback
ISBN 978-0-521-65709-9 Paperback

Transferred to digital printing 2010

For Paul

Contents

General editor's preface

William Schweiker's new book is the sixth in the series *New Studies in Christian Ethics*. Its theme of 'responsibility' is extremely important and apt and has many points of contact with other books in this series. Schweiker argues convincingly that an approach based upon responsibility has much to contribute to the present-day debate about ethics. He also believes that Christian ethics has a distinctive and valuable contribution to make to this approach. Within the secular world an awkward combination of increasing pluralism and technological power makes a notion of responsibility imperative. As power increases in a technological age – so ironically does pluralism. The latter ensures that people become increasingly confused about the bases of morality just at the very moment that they are possessing an unprecedented amount of power. Schweiker, in contrast, argues that an ethical approach based upon responsibility (both individual and corporate), which has moral integrity as its aim, is more apposite.

Drawing on classical and contemporary sources, Schweiker argues that responsibility is linked to our capacity to reflect upon and then revise or transform our lives through criticism of what we care about and thus value. He does not follow those Christian ethicists who have tended to regard responsibility in individualistic terms as a personal revelation or intuition. For Schweiker responsibility involves cognition and critical reflection/interpretation, and is a requirement both for individuals and for moral communities. It is based upon critical reflection aimed at the question of what has constituted our lives under the recognition of and care for others – and, for theists, our lives before God. As he says himself 'conscience is not a faculty of the soul, a divine spark in the mind; it is the practice of radical interpretation within which personal and social identity is constituted and formed in terms of the imperative of responsibility'.

In the final part of this challenging book William Schweiker seeks to show that Christian faith does have a distinctive contribution to make to the general discussion about responsibility. For Christians genuine moral integrity is an indirect consequence of seeking to respect and enhance the integrity of all life before God. An approach which is based simply upon personal autonomy and authentic fulfilment always faces the temptation in a troubled world of the will to power. But for Christians ultimate power is God's alone and faith in this God provides a confidence to live and act amid the fragmentations of life and beyond a culture of personal fulfilment and authenticity. Christian faith offers a vision of 'goodness shining through the fragmentariness and travail of existence, the awareness that being as being is good'. A Christian notion of responsibility is based upon an ultimate power, namely God, who is good, and a finite world that is graciously respected by God.

Responsibility and Christian Ethics has points of contact with several of the other monographs in the series. Kieran Cronin's *Rights and Christian Ethics* repeatedly linked secular language about 'rights' to Christian notions of duty and responsibility. He also argued that Christians have deeper 'justifying reasons for acting morally' than secularists precisely because moral behavior for Christians is a part of their relationship to God. James Mackey's *Power and Christian Ethics* offered a notion of power as moral authority located finally in God which is also very close to William Schweiker's thesis. For Mackey Christian communities at their best offer a 'radical and encompassing sense of life as grace' which 'enlightens and empowers people to imagine and create an ever better life, and also to overcome the forces of destruction which one could otherwise only join and increase, but never beat'.

Schweiker's thesis also overlaps with Ian Markham's *Plurality and Christian Ethics*. Both writers take modern pluralism seriously and Markham, like Schweiker, finally believes that theism offers 'a more coherent description of life than any alternative world perspective – it makes sense of the objectivity of value and the intelligibility of the universe'. For Schweiker the very discipline of Christian ethics is finally 'faith seeking moral understanding'.

By a stroke of good fortune *Responsibility and Christian Ethics* is being published simultaneously with Clinton Gardner's *Justice and Christian Ethics* and both of them quickly follow Jean Porter's *Moral Action and Christian Ethics*. An understanding of Christian ethics similar to Schweiker's is seen in Porter's and Gardner's texts. For Schweiker and these other writers moral action is a product of a subtle and complex juxtaposition of interdependent moral virtues – including, crucially, notions of responsibility. However much one stresses such virtues as autonomy, authentic fulfilment or even justice, a strong notion of responsibility – both individual and corporate – does seem to be required by Christian ethics. Schweiker's book provides the needed theory of responsibility for theological ethics.

I hope that this book gets the serious attention it deserves.

ROBIN GILL

Acknowledgements

Throughout the process of writing this book, I received insight, criticism, and encouragement from friends, colleagues, and students. I cannot adequately express my gratitude to these persons. Still, I wish to thank David E. Klemm, Robin W. Lovin, Douglas Ottati, James M. Gustafson, Kristine A. Culp, Edward Arrington, Franklin I. Gamwell, Terence J. Martin Jr., Susan Schreiner, David Wadner, Kelton Cobb, Jane E. Jadlos, Lois Malcolm, Maria Antonaccio, Darlene Weaver, and James J. Thompson. I also appreciate the encouragement of Robin Gill, editor of New Studies in Christian Ethics, and Alex Wright of Cambridge University Press. The preparation of the text owes a debt to Marsha Peeler, Darlene Weaver, and James J. Thompson. Finally, my special thanks to Kristine A. Culp who read the manuscript and made helpful suggestions on it. Her comments strengthened the book and emboldened my thought. Yet for all of the help I have received, I am alone responsible for the book.

I dedicate this book to my son, Paul. More than anyone else he has taught me that the depths of responsibility and the heights of joy are ultimately one. This insight, I believe, is the truth of our relations to God amid the flurry and fragility of human life.

Introduction

This book is about moral responsibility and Christian ethics. The idea of responsibility is exceedingly complex and relates to all questions in ethics. The onus upon the theologian or philosopher is to distinguish and yet relate the elements of responsibility within ethical reflection. That is what I have attempted to do. The theory of responsibility presented in the following chapters focuses on the relation between value and power in an age of increasing human power. The book seeks to make a contribution to Christian ethics, and also to reflection on responsibility and power in various disciplines.

The book is divided into parts which reflect dimensions of ethics. Part I specifies the *context* for thinking about moral responsibility in terms of beliefs concerning morality and moral agents found in the late-modern Western world. Chapter 1 is an analysis of criticisms of traditional ideas of responsibility in order to clarify values in contemporary life and to show why the idea of responsibility remains indispensable in ethics. This is followed in Chapter 2 with a statement of the main ideas, assumptions, and distinctions of a new Christian ethics of responsibility. The ethic I propose examines the idea of moral integrity and its relation to power and responsibility within moral reflection. These ideas are introduced in Chapter 2, and developed fully later.

Part II presents an integrated *theory* of responsibility. The discourse of responsibility allows us to consider the connection between an idea of morality and our existence as agents. Part II of the book, and thus the theory of responsibility, is divided between these topics. Chapter 3 begins Part II with an analysis of the linguistic and conceptual complexity of the idea of responsibility in moral discourse, and traces the development of this idea in Western ethics. This is followed in Chapter 4 with a typological analysis of theories of responsibility. Within the typology, the chapter also examines particular thinkers whose ideas contribute substantively to the argument of the book.

Chapter 5 presents the theory of value and the imperative of responsibility essential to a Christian ethics of responsibility. This chapter is central to the book, although the theory of value and the imperative of responsibility are elaborated in following chapters. I present a multidimensional theory of value that interrelates goods rooted in human needs. There are naturalistic and also realistic dimensions to this moral theory. This is required in theological ethics because Christian faith insists on the value of created reality and the reality of God as the source of morality. The purpose of the moral life is the realization of diverse potentialities of life. This means that moral reasoning is not solely dependent on the resources of the Christian community, but also on an examination of features of life. On these points the book agrees with so-called natural law ethics. Yet unlike natural law ethics, the position I develop grants descriptive relativism in ethics. "Nature" is not an unambiguous guide to what we ought to do, and morality is bound to other cultural beliefs even though values have a naturalistic cast. The theory of value outlined in Chapter 5 is also an answer to the criticism, explained in Chapter 1, that the demands of responsibility can mutilate human goods.

This moral theory further holds that the basic moral problem of human life is a problem of faith, that is, the problem of what identity-conferring commitment(s) ought to characterize and guide our lives. I argue that moral identity, the coherence of life and its values through time, is bound to the commitment(s) agents hold basic to the meaning and value of their lives. This is why moral integrity, and not only the integration of values in life, is central to the ethics of responsibility. The imperative of responsibility specifies then not only a principle of choice, but also the faith we ought to live by.

The imperative of responsibility is stated thus: *in all actions and relations we are to respect and enhance the integrity of life before God.* Chapter 5 explains this imperative and its relation to the theory of value. What is more, I contend that by acting on this imperative a distinct good, the good of moral integrity, is manifest in human life. By living responsibly, the Christian trusts that the fullness of life is known and experienced. The book as a whole makes this point, marshalls a protest against the moral projects and values which undergird contemporary culture, and proposes an account of the moral meaning of Christian faith. This claim about the good of moral integrity as the meaning of faith completes the discussion of responsibility and morality in the first half of Part II of the book.

In chapters 6–7 the discussion shifts to the question of the nature of moral agents. Throughout these chapters I explore recent work on moral agency and subjectivity by thinkers like Charles Taylor, Susan Wolf, Harry Frankfurt, and others. The purpose of this discussion is to develop an account of moral freedom and identity within an agentic-relational view of persons.[1] Human beings on this view are defined as agents intimately related to each other and their environments. A person's moral being is constituted in and through interactions with others and the critical assessments he or she makes as an agent about life, others, and the world. I develop this argument in Chapter 6 by exploring freedom and moral responsibility with respect to debates in ethics about the relation between freedom and the good. I also show that the ability to have responsibility ascribed to an agent by others and to assume it for oneself is basic to meaningful human life.

Chapter 7 examines the formation of moral identity and thus how we become responsible for ourselves. Drawing on the work of Charles Taylor and others, the chapter argues that a particular interpretive act, what I call the act of radical interpretation, forms moral identity with respect to the diversity of values that permeate life and the imperative of responsibility. The idea of radical interpretation articulates an agentic-relational view of persons within an integrated ethics of responsibility. In fact, it recasts traditional claims about conscience in terms of current work on moral subjectivity and agency. In this way, the realistic and naturalistic dimensions of the theory of value present in Chapter 5 are amended by a hermeneutical account of human existence, that is, an account of how we exist in the world and with others as self-interpreting agents.[2] The resulting moral theory I call hermeneutical realism. In making that argument, Part II of the book is concluded because the connection between morality and agency has been made through the idea of radical interpretation.

Part III of the text explores the *source* of responsibility from a theological perspective. The source of responsibility does not rest on the claim that persons "encounter" God as a personal "Thou," a claim, we will see, basic to previous accounts of responsibility in Christian ethics in this century. The theological dimension, I contend, is specified with respect to the experience of the complexity of value which permeates human life and relations with others and the world. It is this theological claim which demands the new theory of responsibility presented in Part II of the book.

Chapter 8 undertakes the inquiry into the source of responsibility by exploring the recent work on responsibility by Hans Jonas. He argues that the question of power must be central to ethics, and, further, that power makes responsibility morally basic. I show that Jonas' argument raises questions about the source of moral value which can be addressed only from a theological perspective. And, further, I seek to demonstrate the unique claim of Christian faith about God as value creating power with respect to the question of the source of value. Chapter 9 makes a case for the validity of this Christian ethics of responsibility. I thus explain why theological ethics must be understood in terms of an integrated theory. In so far as this is the case, then the ethics of responsibility has been recast beyond the paradigms of thought which have dominated twentieth-century Christian ethics. Part III examines the religious source of the theory developed in Part II even as it answers the root problems endemic to our current situation, problems isolated in Part I of the book. In this way, Part III completes the argument of the book.

Obviously it would be impossible to account for all theories of responsibility in one book. Accordingly, I have tried to isolate basic issues in the ethics of responsibility and engaged those thinkers who, in my judgment, make distinctive contributions to reflection on these matters. In proposing a new Christian ethics of responsibility, I have also not attempted to address all practical matters of responsibility. While some problems in technology, ecology, economic ethics, and theory of punishment will be examined, many questions of responsibility and Christian faith simply cannot be addressed (sexuality, marriage, war, etc.). It is also not the purpose of the book to explore questions about resources in ethics, say the use of scripture in Christian ethics. Finally, my task is not to provide a defense of theistic ethics, although I do sketch a version of the moral proof of the existence of God. My task is to address basic matters in ethics pertaining to moral responsibility to the end of presenting a new ethics of responsibility. The book develops an approach to theological ethics, which, given time and space, could address the range of topics germane to moral inquiry.

A final word of introduction is needed about the enterprise of theological ethics, and, specifically, the term "Christian Ethics." Following an insight of H. Richard Niebuhr, this book is a work of Christian moral philosophy. While I develop a theory of responsibility which differs from Niebuhr's ethics, and, indeed, all Christian accounts of responsibility in this century, it is still one purpose of this book to renew the enterprise of Christian moral philosophy.[3] "Christian ethics" is critical and constructive reflection on moral existence from the perspective of Christian faith. It is also the articulation and even revision of the moral meaning of Christian convictions. Theological ethics is faith seeking moral understanding: *fides quaerens intellectum moralem*.

This book undertakes theological ethical reflection on basic questions of our time. I draw freely from the whole of the Western Christian tradition and from works in philosophical ethics, Jewish thought, and political theory in order to present an ethics of responsibility. Although the point of view in the ethics is Christian, the object of moral reflection is the existence of all human beings. Problems of responsibility are not unique to Christians. Whatever else we are, we are all members of one species sharing a fragile planet. All forms of ethics must address this simple but important fact.

I

The context of responsibility

CHAPTER I

Responsibility and moral confusion

Questions of responsibility are as old as human civilization itself. The myths, symbols, and rituals of the world's religions, archeological evidence from early human communities, and all forms of social life testify that human beings have always engaged in acts of praise and blame, debt and reparation, making obligations and fulfilling or failing to keep them, questioning and answering, acting faithfully and without faithfulness. These actions and social practices are the simplest and most pervasive forms of responsibility. On the basis of historical and sociological evidence, one might conclude that to be human is to be responsible. Debates about the grounds, norms, and limits of responsibility are at root disputes about the meaning of our humanity and our lives together.

Contemporary Western societies are riddled with unending criticism of traditional moral convictions, and, therefore, conflicts over the meaning of human existence. At the same time, there is within these societies an increasing demand for persons and institutions to assume responsibility for all domains of life: family, politics, economic life, medicine, and a livable global environment. We can begin the inquiry into responsibility and Christian ethics by exploring criticisms of the traditional ideas of morality and responsibility. This is important since many, but not all, of these ideas about morality have roots in Christian faith. Later in this chapter I will relate these criticisms and the values behind them to problems which make responsibility morally central in our time, especially the fact of moral pluralism and the reality of human power. The way forward in ethics requires a reconstruction of responsibility which takes these criticisms and problems into account.

9

RESPONSIBILITY AND CONTEMPORARY MORAL VALUES

Criticisms of the idea of responsibility center on a claim found in much of Western ethics and virtually all of traditional Christian ethics. What is under criticism is the belief that the consideration of the well-being of others or one's duty to God ought to determine a person's conduct and also what kind of life he or she should strive to live. Morality is defined by obligation to others which includes reasons for self-sacrifice. William K. Frankena has clearly delineated this conception of morality. He writes:

morality is a normative system in which evaluative judgments of some sort are made, more or less consciously, from a certain point of view, namely, from the point of view of a consideration of the effects of actions, motives, traits, etc., on the lives of persons or sentient beings as such, including the lives of others besides the person acting, being judged, or judging (as the case may be.)[1]

The moral point of view requires the consideration of the lives of others in deciding what one ought to be and to do. I should do unto others as I would have them do unto me. We are to love others as ourselves (cf. Matt. 19:19; Mark 12:31; Luke 10:27; Rom. 13:9; Gal. 5:14; Jas. 2:8). Christian faith intensifies the principles of moral equality and reciprocity through its conception of love, or *agape*. The Christian is to act out of radical other-regard, a revolutionary love for others and their well-being beyond equal regard.[2] We are to love our enemies. The logic of equal regard, or reciprocity, is transformed in Christian ethics by the reality of superabundant love.[3]

Critics point out that the demand for impartial other-regard, let alone Christian *agape*, as definitive of morality can mutilate genuine human goods.[4] It seems to require that we view our own lives from a disinterested perspective and thus forsake the commitments, beliefs, and projects that make us who we are. And yet it is precisely those commitments which provide a framework for making sense of life. The idea that life is lived under obligation to others can stunt human aspirations and also establish intricate systems of debt, retribution, and guilt, systems found in virtually all cultures. The criticism is significant, because it tells us something about the problems contemporary ethics faces in thinking about responsibility.

Many people live with a crushing sense of responsibility. These persons sacrifice unduly their own aspirations out of a sense of duty to others, or they live with guilt for things they could not have reasonably altered or avoided. Feminist ethicists, for example, argue that the "sin" of women is not prideful self-aggrandizement but suppression of their needs, sensibilities, and actions to the demands of traditional roles and obligations to others. Service to others is claimed to be the path to fulfillment, even if, we should note, the demands for service are not equally distributed among persons. The traditional conception of morality has been used to legitimate "feminine virtues" which actually perpetuate oppression. As one author notes:

> We appeal to altruism, to self-sacrifice, and in general, to feminine virtuousness in a desperate attempt to find grace and goodness within a system marked by greed and fear. However, while these virtues may herald for us the possibility of ethics – the possibility of some goodness in an otherwise nasty world – nevertheless ... they are the virtues of subservience.[5]

Morality, the criticism goes, is destructive of genuine human aspirations or is merely the tool of the powerful to oppress others. Oppressed persons learn to understand and value their lives in terms of the dominant social system, a system which does not work to eliminate oppressive structures.[6] The psychological power of obligation shapes and stunts self-consciousness and thus an understanding of human potential. If we could drop the overriding demand for impartial other-regard as definitive of what is morally praiseworthy, then, the critic holds, we might live less guilty, less stunted, and genuinely happier lives. One could grant obligation a place in human life but not see it as definitive of the morally good life.

This criticism of commonly held notions of obligation and responsibility as well as the Christian conception of neighbor love expresses a pervasive moral outlook. The basic conviction found in contemporary Western cultures is that human life ought to be characterized by the search for fulfillment and authenticity.[7] Fulfillment is not defined in terms of obedience to social roles, cultural ideals, or the perfection of a certain set of virtues. It is defined with respect to enhancing the richness and complexity of a person's life. Similarly, authenticity is understood not in terms of fidelity to objective moral standards, values, or codes of behavior, but, rather, in terms of being true to oneself and one's basic commitments. The critics of traditional morality are arguing, then, that individuals ought to seek fulfillment and be true to themselves, be authentic. In so far as responsibility seems to demand some sacrifice of the goods one wants and values, it might be best, on this account, simply to demote its importance in understanding personal life. A basic feature of our situation is, then, that the values of fulfillment and authenticity provide the means to assess the demands of moral obligation. In this way the contemporary moral outlook diverges from traditional morality and Christian love.

However, the criticism of traditional morality centers not only on ideals of human excellence. Since it expresses beliefs for orienting human life, the criticism also focuses on human capacities for action and beliefs about the world in which we live. In this light, traditional discourse about moral responsibility backed the belief that the world is open to moral evaluation. Examples of this conviction abound in the ancient world and continue to shape the moral imagination of the West. For instance, in the book of Job there is an assumption that some agent other than human beings is involved in Job's suffering. Moreover, the backing for specific commandments in the Bible is what God has done in history, for example, how God liberated Israel from bondage (Deut. 6:20–25). It is essential to the Bible that God is an agent in human affairs. Goods and obligations are constituted by God's action even as reality itself is open to moral assessment. These beliefs are not unique to the biblical world. They dominated Greek and Roman thought as well. Sophocles' powerful drama *Antigone* revolves around the obligation Antigone has to her brother under the law of the gods. The Sophoclean vision traces the tragic working of fate in and through human affairs. Forces other than human beings have moral meaning. In each of these cases, and other examples could be used, the belief is that reality is open to moral evaluation because human beings are not the only morally relevant agents in the universe.

The conviction that the world, fate, or the divine is acting for or against persons and that this "action" is open to moral assessment continues to grip the imagination. As James M. Gustafson has argued, human beings have a sense of powers bearing down upon them and also sustaining them.[8] It is important how these "powers" are understood simply because that interpretation shapes the way in which a person or community acts in the world. For example, if we see the natural world as something to control, we act differently than if we behold it with reverence. Similarly, there are situations in which persons desperately seek to find someone or something responsible for what has or will happen to them, but, alas, can find no one to blame. The patient with a terminal disease wants to know why this is happening to her. But there is no simple answer, no one person or thing to blame. Even to blame God is self-defeating since it reveals to the blamer that God is unworthy of faith and love and thus not really God. It exempts God from responsibility by practically disproving the validity of belief in God.

It is at this juncture that the critic of traditional ideas of responsibility presses a question about the character of moral worldviews, that is, metaphysical beliefs. While understandable, the complaint of the person in pain requires that we accept the assumption that the discourse of responsibility is somehow adequate for understanding the way the world goes. That assumption must be questioned and finally rejected. It might help to view nature reverently, but in no sense ought we to assume that "nature" is an agent or that our "view" hooks into reality. We simply have no evidence that reality, nature, or God can be held morally accountable in the same way human agents can. The critics contend that if we were to excise traditional ideas of responsibility from our lives, we could then rid ourselves of the pain that accompanies unanswered questions and also false beliefs about the world and ourselves (like belief in God or that the world is punishing us).

The criticism of traditional morality and Christian love centers then on ideals of human excellence and how beliefs about responsibility inform our understanding of ourselves and the world. Moreover, the criticism is grounded in the contemporary values of authenticity and human fulfillment. The quest for freedom from guilt, deception, false belief, and unduly burdensome obligations seems to require an escape from responsibility as it has been understood in much of Western ethics, especially Christian ethics. Traditional conceptions of morality as well as the discourse of responsibility seem at odds with the deepest aspirations of contemporary persons. Can we show why one must reconstruct beliefs about responsibility rather than simply dropping them from ethics?

The next two sections of this chapter are dedicated to that question. Let us begin by trying to imagine life without responsibility. If this should prove impossible, then it would appear that this idea is indispensable to moral thinking. In showing this, we must focus on the concerns of the critics of responsibility, namely, the meaning of personal life and also the place of moral beliefs in understanding the world.

RESPONSIBILITY AND THE WORLD OF AGENTS

In order to picture human life without responsibility we would have to conceive of individuals stripped of the language, values, and attitudes which are the condition for their interactions with others. An individual, as Schubert Ogden notes, is "a center of interaction that both acts on itself and others and is acted on by them."[9] Without some idea of responsibility these individuals would lack real self-understanding, since each would have no means for designating herself or himself as an agent who interacts with others. We could never answer the question "who is responsible?" with respect to our actions or anybody else's. These individuals would show no trace in their lives of the decisions they made or the reasons for their actions. We could not really call such beings "selves" in any profound sense of the term simply because it would be inappropriate to ascribe to them the choices for their actions. Their lives would be immersed in an immediacy of thinking, willing, or feeling because no one could designate how a past "self" relates to a present "self" with respect to its choices, relations, and future projects. Memory and hope would be ruled out for individuals; authenticity and fulfillment could hardly characterize their lives.

A society of these "individuals" would be a collection of human beings with no means to think, speak, or evaluate their interactions with each other. It is debatable whether or not we can really conceive of them forming a society even on the basis of brute self-interest, in spite of what theorists from Thomas Hobbes onward have held. Self-interest requires some connection between one's actions and one's desires and sense of identity. Lacking that connection, we would never know when an action attained the end "we" wanted. Yet the link between action and agent is precisely what the idea of responsibility articulates. To excise the idea of responsibility from the center of ethics is then to forsake the conditions necessary to be genuine selves and also to have relations with others. This might indeed be a life without guilt, burdensome obligation, and possible deception or false hope. But it would also be a life without praise, joy, admiration, indignation, anger, regret, gratitude, memory, or genuine hope.

Our lives are enmeshed in patterns of responsibility. This is because human life is irreducibly social. These patterns of interaction are the necessary condition for living any form of human life as we presently know it and can honestly imagine it. What is more, assigning responsibility to an agent, or to some community of agents, focuses moral evaluation on actions and also on the character of the persons or community involved. We want to know "who" is responsible and not simply what action or series of events transpired. Responsibility would seem, then, to be central to any quest for self-understanding. We understand ourselves as persons to the extent that we and others can assign some identity to ourselves as actors.

This fact is also important for how we conceive of the world. A world without responsibility would be a world without agents, a world of sheer events and happenings. To be sure, we might want to question, in fact we must question, appeals to non-human agents in the world. But that question presupposes that the world is a place of agents. The dispute is simply over what kind of agents are acting in the world. The idea of responsibility is important, then, because it means that an account of reality, that is, metaphysical beliefs, must make sense of the fact that there are agents who act and suffer and make choices about how to live. This is basic to moral self-understanding simply because without the idea of responsibility, or something very much like it, we would lack the means to talk about ourselves as agents intimately related to others and our environment. We can clarify this point by considering some real, if painful, facts of contemporary life.

Nearly every week in the city of Chicago at least one child, one woman, or one man is killed by gang violence or physical abuse. This is true in major metropolitan areas around the globe. The figures of violence against persons are higher if we include death by starvation, the horrific consequences of girls and boys forced into prostitution, the worldwide slave trade, various forms of infanticide, and warfare. Now, the parents of a child who is murdered by gang violence do not want to know simply about the gun or the bullet. They want to know who is responsible for the death of their child. The parents want to understand the intentions and reasons for the killer's actions. Why did this person kill my child? And the same question is germane when a child is abused or sold into prostitution or when a human life is violated. Why was this done?

These painful facts expose a set of problems for ethics which center on the idea of causality, and thus, once again, on beliefs about the place of responsibility in understanding the world. It is the case that the gun and the bullet were in some sense the cause of the child's death by the gang member. Men and women use horrific means in the abuse of their children. The instruments of abuse and murder are all in some sense "responsible" for the harm done to persons. If we were to give a purely physicalist account of what happened, the focus of attention would be on these factors. But, within the series of events leading to the murder or to other violations of human life, one actually seeks in moral reflection to isolate an agent who exercises power in order thereby to understand and evaluate what happened in moral as well as causal terms. We seek not only to explain what happened, but also to evaluate it, to praise and blame those involved. The way the world goes from a moral point of view entails different kinds of judgments about causes in the world.

The discourse of responsibility involves both causal and evaluative judgments in trying to specify valid acts of praise and blame. As Marion Smiley has pointed out, the conceptual problem is to clarify the connection between these two kinds of judgments.[10] Is an evaluative judgment about responsibility valid only if it is made with respect to someone who directly caused something to happen or failed to act? We would think it wrong that a fifteen-year-old boy was blamed and held in custody for the death of a member of a rival gang if there was not reasonable suspicion that he directly caused the death of that person. In fact, we would judge the boy innocent if it could be shown that he was not the cause of the death. The intuition here is that agents "own" their actions and thus are responsible for them. Given a certain act, we try to establish who "owns" that action, who is responsible for it. And yet granting this point, it is still the case that causal and evaluative judgments are not identical. If they were identical, then it would be nonsense to hold other gang members or members of a family responsible in any degree whatsoever for the violence one gang member commits or a parent inflicts on a child. That conclusion cannot possibly be right. Complicity is also a form of responsibility.

Theorists of responsibility differ on how to address these matters. Without entering those disputes at this point, we must grant for the moment the connection between causal and moral responsibility because it is important for understanding the contemporary moral outlook. This outlook insists that it is proper to understand events in moral terms only if it is possible to isolate an agent who caused or refrained from causing something to happen. Moral responsibility is only properly assigned to persons (not gods, human collectives, natural events), and the ability to be held responsible is a defining characteristic of what it means to be human. This is why the patient with a terminal illness who asks "why me?" can find no answer, since it is difficult, if not impossible, to isolate a single agent in the stream of events which is her disease. It is also why we make a categorical distinction between the responsibility of the gang member and the "responsibility" of the bullet for the death of his victim. The contemporary outlook is dominated, then, by the supposition that human beings alone can properly be called agents and that the world in which we exist must be described in non-agential terms. The world is not explainable by appealing to supra-human agencies even as we seek to understand the world with respect to the possibilities of human action. The depth of this supposition about the uniqueness of persons in a non-agential or non-personal universe can hardly be doubted. As social commentators have pointed out, modern persons live in a disenchanted universe. The insight is that we must isolate an agent in the flow of causal events if we are to understand those events morally. This requires persons to face the gravity of their existence as agents; it forestalls appeals to forces which might mitigate the burden of responsibility for actions. In order to reconstruct our ideas of responsibility, we must, given this outlook, explore the connection between causal and evaluative judgments. And we also need to explore how responsibility is assigned to agents in terms of those judgments.

The connection between causal and evaluative judgments and the assignment of responsibility is easily grasped. We can illustrate this by recalling the story of the Garden of Eden (Gen. 3:1–24). At one point in the narrative God asks about Adam's knowledge of his nakedness. Adam rightly takes this as a causal and an evaluative question; it is about what was done and the rightness or wrongess of the act. But interestingly that connection does not determine how Adam tries to assign and evade responsibility. Adam attempts to shift the burden of responsibility, and thus the guilt for his disobedience, to someone else by insisting that he was not the cause of thinking about eating the "apple." He blames Eve. She gave him the apple and then he ate. The phenomenon of the shifting of blame is important, as scholars point out, for social practices such as scapegoating.[11] Although it is not often noted, the shifting of responsibility can surround good actions as well. It is difficult, if not impossible, to take credit for being the sole cause of good things that happen.[12] But the point of God's question to Adam was not about the causal origins of thoughts. The question was why Adam in fact so acted; it was about the morality of a choice. The problem of shifting responsibility seems to force theorists to the conclusion that proof of direct causal responsibility must undergird evaluative judgments in assigning responsibility.

Thus, in matters of responsibility we want to know who is acting and to evaluate actions in moral terms because of the assumption that there are connections between the self-understanding of an agent (Adam), or community of agents, their actions (eating the apple), and some framework of values used to orient life (God's command). This tells us something more about the nature of moral agents. A moral agent, as opposed to other forces in the world, is a fit candidate for attitudes like commendation, anger, respect, indignation, gratitude, love and so forth. In the Genesis story, Adam and Eve are open to these kinds of attitudes. They are punished by being exiled from the garden and thereby have their deep estrangement from God made spatially manifest in the story. Even the serpent is condemned to slither upon the earth. Similarly, the gang member who murders someone is subject to indignation and punishment, or, astonishingly, perhaps admiration from his fellow gangbangers. In each case, the identity and self-understanding of an individual (Adam, Eve, the serpent, the murderer) is bound to the assignment of responsibility because the person's life is interpreted with respect to an awareness of the gravity of being a moral agent.

However, there are important conditions that surround the idea of moral agency in speaking about individuals. It is one of the most protected concepts in moral discourse. In most cultures high standards must be met before one is entitled to be called a moral agent simply because this designation entails membership in the moral community and thus the right to make claims against that community with respect to one's well-being. Within modern societies a crucial standard for membership in the moral community is that an individual be in control of his or her life in some important respect. This requires a measure of freedom on the part of the agent as well as some capacity to intend and deliberate about courses of action. For this reason animals are often excluded from the moral community, and, in patriarchal cultures, women and children, thought to be less rational than males, have

at best second-class status in the community. In the West, moral
innovation has come only when those excluded from moral
standing demand their rights in terms of standards for assigning
responsibility, or when criteria other than freedom and rationality, as, for instance, the simple capacity to feel pain, are taken as
the baseline for moral standing. In this respect, the idea of
responsibility is a clue to a deep problem in contemporary life.
What are the standards for admission to the moral community,
especially when previous standards have proven unduly exclu-
sive?

Traditionally, responsibility ethics takes agency rather than the
capacity to feel pain as crucial for admission to the moral
community. Without some measure of control over their behavior, agents would be inappropriate candidates for certain attitudes associated with responsibility. The value contemporary
Western societies place on human power and also on authenticity
and fulfillment is characteristic of the operative standard for
admission to the moral community. And this is why, as liberation
and feminist theologians note, moral innovation in these societies
requires the empowerment of persons who have been excluded
from active participation in shaping their world to be historical
agents through increased self-understanding of their own power
to act and change the world.

The importance of control over one's life as a condition for
assigning responsibility is, of course, not new. It is why the story
of Adam and Eve centers on God's command and also on the
reality of temptation and choice. The background assumption of
this narrative is that Adam and Eve have some control over their
lives. Likewise, the gangbanger or abusive parent might plead
insanity in order to avoid or lessen responsibility for a child's
death or they might appeal to their own experience of being
abused. But the success of the plea depends on showing that the
individual is not an appropriate candidate for some fundamental
attitudes we normally hold about persons. The plea of insanity is
meant to lessen the demand of living in recognizably human
ways, since, if proven true, it means that the person had no
control over her or his life. Ironically, this shows that taking
responsibility is basic to authentic human life.

The inquiry of this chapter is revealing that the discourse of responsibility bears on how persons and societies think about what it means to be human and the significance of life. And it implies a view of the world and social life in which it is possible and proper to assign responsibility to agents because they are in some measure in control of their actions. As Wilhelm Weischedel put it, "responsibility thus presupposes that a person is not determined to act by another, but, rather, is the cause of his or her action."[13] Little wonder that an ethics must examine and articulate what underlies the idea of responsibility, the reality of human moral existence and the possibility for action in the world. The discourse of responsibility provides the means to debate the meaning of human life and also beliefs about the world. This is why it is central to all current moral disputes.

RESPONSIBILITY AND PERVASIVE MORAL PROBLEMS

I have been arguing that ethics must demonstrate how the idea of responsibility enables us to articulate and evaluate widely held moral beliefs. By exploring the moral landscape of contemporary culture, I have also tried to show why the idea of responsibility is needed in ethics. The task now is to connect that discussion with pervasive moral problems. In doing so, I want to show that the values of fulfillment and authenticity by themselves lead to moral confusion. If that can be established, then I will have cleared the way for the remainder of this book.

The problem of moral pluralism

The first problem which all contemporary ethics must face is the fact of moral pluralism. Yet in one respect, the reality of pluralism is unremarkable. If pluralism is synonymous with "plurality," then it is simply the case that there have always been diverse moral outlooks.[14] However, the term pluralism means more than simply the recognition of a diversity of moral outlooks. It is the view which affirms that plurality is a good thing and argues that this affirmation must be constitutive of any valid moral outlook.

Pluralism as a moral conviction raises problems for ethics. The

basic one is the possibility that moral outlooks are incommensur-
able, that is, that one cannot translate a set of moral beliefs into
another, develop a neutral moral language, what Jeffrey Stout has
called "moral Esperanto," or specify a common morality such
that a viable comparison and testing of beliefs can take place.[15] In
so far as moral beliefs are the condition for our thinking,
speaking, and acting together, then the fact of incommensur-
ability would mean that we have no way to understand those who
hold moral outlooks different than our own. If this is the case,
then it is impossible to make moral judgments about the beliefs
and actions of persons and communities holding different moral
outlooks. We would lack the cognitive conditions for understanding others. This also raises suspicions about one's own moral
beliefs, that one might be unintelligible to others. Is it possible to
make moral judgments across cultural boundaries?[16]

A negative answer to this question can be formulated in two
ways. One version, let us call it soft relativism, contends that
moral principles are valid only for the cultures in which they are
found. In this form of relativism, there are distinctive moral
factors (i.e., moral values and norms not reducible to other facets
of life), but these are interrelated with other factors in a culture,
like beliefs about family structure, political power, or human
finitude. Moral factors vary with these other cultural beliefs and
practices. The truth of moral principles is determined with
respect to the beliefs about the world and human life represented
in different cultural traditions. This form of relativism is "soft" in
the sense that it does not deny that moral beliefs are distinctive
and can be validated or shown to be true. Yet the validity of those
beliefs and the practices of validation are relative to other factors
in a culture.

Hard relativism insists that no moral beliefs can be shown to be
true or false. In this form of relativism "moral principles are
causally dependent on something else, e.g. inter-familial relation-
ships, or early training within a particular culture, or on the form
taken by the economic or power structure of a society at any
time."[17] There are no distinctive moral factors because all moral
beliefs, like those about responsibility, can be reduced to non-
moral conditions or relations. And since there are no distinctive

moral factors, a particular moral worldview makes no claim to be true, but is simply an expression of personal preferences or a matter of conventional mores for guiding human action.[18] For instance, if one finds the Inuit, or Eskimo, practice of abandoning the infirm morally abhorrent, this is simply because one's sensibilities have been formed in a certain way, or, what is really the same thing, we have created our identities from the available moral lexicon. One cannot make judgments about the moral conduct of other cultures since on this view moral beliefs cannot be validated. The difference between hard and soft relativism, then, is that relativism in its hard form sees morality as the result of other things while in its softer form there are distinctive moral elements in a culture which are interdependent with other aspects of the culture.

The other answer to the problem of moral pluralism is to contend that it is possible to make judgments across cultural boundaries. As Dorothy Emmet points out, it is one thing to argue that it is descriptively the case that moral convictions vary between cultures and historical periods. Yet it is another thing to hold, as hard relativists do, that the rightness or wrongness of actions and the goodness or badness of ways of life are themselves relative.[19] The contention of this book is that the human condition sets limits to the forms that any way of life can take, the scope of moral concern, the roots of moral motivation, and even the social effectiveness of action. As Patricia Derian, an official in the Carter Administration, noted, "Electrodes applied to the gums shatter teeth in the same way in Manila as in Moscow. Cruelty knows no [distinctions] ... the pain is universal, the demeaning and degrading of individuals is as hateful to those in the People's Republic of China as it is in South Korea."[20] If this point is granted, it would not deny descriptive relativism, but it would call into question the intelligibility of the idea of normative relativism. Without extremely strong counter-arguments some things are wrong simply because they violate basic features of existence. In an age of global responsibility, the possibility of judgments across moral boundaries must be established and defended.

The ethics of responsibility developed by theologians in the mid-twentieth century did not address these matters of pluralism. Some theologians, like H. Richard Niebuhr, thought that the diversity of cultural beliefs challenged the radical monotheism of Christian faith. The possible validity of other moral beliefs did not enter into consideration in developing an account of responsibility. Other theologians insisted on obedience to the Word of God. Outside of the command of God ethics is sin, according to Karl Barth. Many Roman Catholic theologians relied on transcendental arguments, that is, that all persons, whether they know it or not, always and already have some orientation to God. Moral diversity was then not a problem. The theological approaches to responsibility earlier in this century did not internalize the problem of pluralism. This fact renders these forms of ethics practically inadequate no matter what gains they might have represented within Christian theology.

The burden of this book is to show that an integrated theory of responsibility provides a framework within which moral debate can take place. I call this position hermeneutical realism in ethics. It grants descriptive relativism with respect to how moral values are understood and interpreted; it insists on historical and pluralist consciousness in ethics. Yet I want to show that this does not entail normative relativism with respect to the diversity of goods rooted in created life. The way forward, I argue in Chapter 5, is through a multidimensional theory of value rooted in natural needs and relations. This brings us to the second pervasive moral problem of our time and also to the need for an imperative of responsibility.

Questions of power

From the perspective of an ethics of responsibility, the most pressing moral problem is the radical extension of human power in the contemporary world. This power is manifest in communication systems, economic interdependence, the environmental crisis, and the threat of mass destruction. It has made our planet into a global village composed of wildly diverse moral communities. The pressing nature of the question of power is not hard to grasp. For example, what does it mean in medical ethics to speak of responsibility when through genetic engineering we now have the capacity to alter future generations of the human species? This fact might mean, as Paul Ramsey pointed out, that those who come after us will not be like us.[21] What responsibility do we have for future members of a species genetically different from us? Can we be held accountable for actions the effects of which will not be known for generations? If we *can* develop these technologies, *ought* we to do so? Technology so extends human power that future life is subject to human power and is, therefore, also our responsibility. Much current ethics centers on the radical extension of human power through technology in all of its forms, for instance, medical, military, communicational, and environmental technology, and then examines the demand for responsibility. An ethics of responsibility is about the evaluation and direction of power.

Power can be understood in a variety of ways.[22] Conceptions of power range from understanding it ontologically, that is, as a claim about being, to political definitions of authority and ideas about charismatic leadership. Some scholars distinguish domination, or "power-over," from patterns of mutual empowerment, or "power-with," and the ability to act or be a cause in the world, "power-to." Currently there is intense interest in the dispersion of power through networks of social relations. Obviously an extensive examination of the relation between power and responsibility would have to consider these conceptions of power and the theories used to specify them. For the sake of my argument, it is best to use the most basic definition of power. Power is the ability to produce effects in the world and thus the force or energy to act. As the feminist ethicist Beverly Wildung Harrison notes, power "is the ability to act on and effectively shape the world around us, particularly through collective action and institutional policy."[23] Power can be exercised, shared, evaluated, and justified in a number of ways, as various theories of power in fact show. An ethics of responsibility is concerned with the way in which the increase in the capacity to act and produce effects in the world raises problems of specific interest for human conduct.

In one respect, the relation between power and human action seems obvious. If an agent is powerless to act, he or she cannot conduct personal life by norms and values which are to determine choices about actions or what kind of person to be. Normally persons are not responsible if they lack the capacity to act, are coerced into acting, or their movements are involuntary. Similarly, because many actions rely on and require the cooperation of others, there are social conditions which rightly figure in assigning personal responsibility. Our current situation is characterized by the radical extension of human power. Persons in advanced technological societies now have capacities for action previously unknown in the history of the world. This increases the degree of responsibility these persons, institutions, and even societies bear for the viability of life on this planet. But it also means that individual action is vulnerable to institutional and technological forces beyond personal control. If an individual decides, for example, that she or he should support relief measures for starving children, then the effectiveness of her or his intentions requires cooperation with others. One might contribute to the Christian Children's Fund. But in that case, one's voluntary action is rendered dependent on the purposes and responsibility of that organization. Conversely, persons excluded from participation in social and economic institutions are increasingly disempowered and are deprived of responsibility for the course of their lives and communities. We are witnessing a radical differentiation of responsibility due to the imbalance of access to the institutional means of power. In this situation those without access to power and barred from institutional structures often bear disproportionate burdens for the consequences of social and economic action for which they are not responsible. The increase of human power in our time is then in good part a matter of institutional developments.

I will return to these matters later in this book. At this point we must ask about the moral value of power. In general, thinkers have valued power in two ways. First, they have argued that the value of power is found in attaining certain ends. The basic pattern of thought, as Alasdair MacIntyre has argued, entails some conception of what persons in fact are, and also what persons ought to become.[24] The moral life is the movement from what one is to attaining some ideal of human flourishing. We might imagine different ends of human life. A Christian theologian, like Thomas Aquinas, argues that the human good is found in God; a utilitarian insists that we ought to seek the greatest good for the greatest number of sentient creatures. Yet in each case, the meaning of the human power to act is understood and evaluated in terms of the condition, state of affairs, or end sought and actualized. This outlook deep in Western consciousness finds expression in the contemporary moral ideal of fulfillment. Importantly, this value involves an assessment of power. It provides a principle of choice for how to exercise power with respect to other values.

Second, philosophers and theologians have argued that the moral worth of human power is not what is attained through its exercise, but in the act itself. There are certain things which are to be respected in all actions no matter what actual ends are sought or realized. Persons can seek any number of ends, but the exercise of power is morally good if and only if it respects certain inviolabilities. The value of an agent's power to act is determined by the norms to be affirmed in the exercise of power. This idea finds expression in contemporary life in terms of the ideal of authenticity, itself an evaluation of power. In all action we ought to be true to ourselves, which means never to forego the quest for self-fulfillment. This again is a principle for determining how to use power with respect to other values. Fulfillment, as an end or good, and authenticity, as duty to self, provide, as we saw before, an *ad hoc* moral framework of value for understanding and evaluating human power.

These diverse evaluations of power are a clue to our current situation. Advanced technological cultures are riddled with moral confusion and uncertainty. This is due in part to the reality of moral pluralism in an age of global interdependence. But it is also so because the increase in human power and the institutional mediation of human action challenges the extent to which we can in fact direct the exercise of power. The radical extension of human power in our time threatens to overwhelm moral reason, making all moral reflection instrumental to the simple purpose of furthering human power. What is threatened by the radical extension of human power is the very framework of value in which we make moral sense of our lives. At a personal level this is reflected in the fact that the idea of authenticity is defined in these cultures in terms of self-fulfillment. Truthfulness to self places no constraints on the self and the quest for power since the concept of authenticity introduces no new commitments or values beyond that of fulfillment. As we will see in the following chapters, it is at this level that the idea of moral integrity becomes important in ethics.

The importance of responsibility

Once again in relation to questions of pluralism and power, we see the importance of reconstructing beliefs about responsibility within ethics. In so far as responsibility expresses the connection between agent and deed, it is a crucial concept for an age in which the human capacity to intervene in the world is increasing. Similarly, in so far as the idea of responsibility implies that an agent or community of agents is responding to others, then this idea is also helpful for moral thinking in a pluralistic age. The idea of responsibility seems to provide the means for thinking ethically in an age characterized by moral diversity and the increase in human power. Much as we cannot excise responsibility from ethics if we want to make sense of contemporary ideas about human agency and the world, so too responsibility is crucial to any ethics that wants to address the actual problems persons and societies face. In this book, I develop a theory of value to address the problem of moral diversity and also an imperative of responsibility which is to direct the exercise of power.

RESPONSIBILITY AND MORAL CONFUSION

I have now isolated a variety of reasons why responsibility matters in the moral outlook found in contemporary Western societies. There is a pervasive desire to rid ourselves of inappropriate feelings of guilt as well as to overcome the forms of self-deception and false beliefs that too often surround the moral life. The moral ideals of fulfillment and authenticity often seem at odds with the demands of responsibility. Life might be easier and perhaps less conflict-ridden if we were to ease these demands on us. Yet, I have shown that the beliefs to which persons appeal in lessening the demands of responsibility are deeply moral in character. Precisely because that is the case, the criticism of responsibility cannot be the whole story about the moral life.

Given this, I have also argued that hidden in the idea of moral responsibility is a pervasive moral outlook that actually sustains the suspicion of inordinate moral demands. By bringing this outlook to light, we have seen that it is characterized by an insistence on the reality of human power to intervene and shape the world, a naturalized view of reality, the awareness of the ambiguity of moral judgments, a concern for authenticity and fulfillment in life, and also the question of who is an appropriate member of the moral community and thus can rightly make a claim for our recognition, respect, and concern. Making a case for an ethics of responsibility must address these various features of the late-modern moral outlook. And this is certainly true for a theological ethics of responsibility, since, as noted, many contemporary assumptions about responsibility challenge previous religious convictions.

We have actually reached a startling conclusion. In late-modern Western cultures responsibility is criticized for placing inordinate demands on persons. Yet, the idea of responsibility is also essential for contemporary beliefs about human life and the world in which we live. Oddly enough, contemporary social life requires belief in responsibility in order to sustain its view of the world and yet must deny it with respect to basic orienting values. Here is, I judge, the root of moral confusion in current cultures. Beliefs about the nature of agents and the world in which they live and act are inconsistent with the values that persons believe ought to guide their lives. In this situation, we might expect to find a culture in which the pursuit of fulfillment and authenticity is disconnected from obligations that bind persons in virtue of their lives together. We might expect to find a culture in which responsible existence is perceived as a necessary evil of social life but not essential to human well-being. We might expect to find a culture in which persons attempt to shift responsibility for their lives in the pursuit of what is perceived as fulfilling and authentic and in doing so willingly deny their dignity as agents in the world. We might expect to find a culture in which responsibility for future life on this planet cannot motivate present action and policy. In other words, we might expect to find the culture in which we live.

The only way forward in ethics is to reconstruct our discourse of responsibility realizing that this entails a transvaluation of values. The remainder of this book is dedicated to that task. In the process of carrying it out we will see that the values of fulfillment and authenticity must be transformed in terms of moral integrity. What is more, theological discourse, I will show, is essential to making sense of the idea of moral integrity and thus a coherent theory of responsibility. In the process of making this argument I will confirm the point of this chapter. The idea of responsibility expresses *in nuce* an entire moral worldview. This worldview is one in which the power and dignity of human agents ought to respect and enhance the integrity of life. It is a worldview which is intelligible once all things are understood and evaluated as existing before God.

CONCLUSION

In this chapter I have tried to paint in broad strokes the moral landscape of contemporary Western culture. In the following chapters this picture will be given greater detail and texture. But at this juncture it is enough to have shown that moral beliefs embedded in these societies, beliefs about fulfillment, authenticity, and responsibility, are at odds with each other. Given this, these societies are unable to address the problems posed by the reality of human power. This fact requires a cogent theory of responsibility to aid us in how to live.

CHAPTER 2

A new ethics of responsibility

In the previous chapter I argued that the problem facing culturally diverse and technologically advanced societies is confusion about which values, norms, and beliefs ought to guide our lives at the very moment when human power is expanding radically and in previously unknown ways. The conjunction of this crisis of values with the extension of power creates a situation in which political, economic, moral, and religious ideologies champion the exercise and pursuit of power as the meaning of life. From a Christian perspective, to make power the supreme value of life is to deny and violate the meaning of being human. It is idolatry to worship as the supreme good something other than God (e.g., Exod. 20:1–3; Matt. 19:17). For Christian faith, the empowerment of persons to be responsible agents in history is to serve the purpose of respecting and enhancing the integrity of life before God.

The last chapter also led to the conclusion that it is necessary to reconstruct the ethics of responsibility. In order to do so, I will develop an ethics of responsibility by interpreting the resources of the Christian tradition in critical engagement with other forms of thought. The purpose of this undertaking is to understand the meaning of moral responsibility and human life from a theological point of view. In this chapter I want to introduce the basic ideas and distinctions of an integrated ethics of responsibility.

MATTERS OF RESPONSIBILITY

Any ethics of responsibility proposes an account of how we ought to understand moral responsibility and what this means for how we can and should live. These are large and difficult matters. The complexity of these matters is due partly to the theoretical issues involved and partly to the fact that questions about responsibility touch virtually all of the practical problems late-modern, technological societies face. Not only must the moral theorist explore what it means to be responsible, but the scope of human responsibility now reaches into the genetic structure of life and the future viability of life on this planet.[1] An ethics of responsibility specifies the moral evaluation of the exercise of power by agents or communities of agents. Matters of responsibility are matters of the use of power by agents who act and suffer. An ethics of responsibility seeks to determine the morally proper use of power. Yet when one turns to works on responsibility, one finds a confusing, even conflicting, array of theories. It is not at all clear that "responsibility" means the same thing in each theory. There may really be no common characteristic but only a family resemblance among theories in the use of the term responsibility. It is my intention to draw insights from various theories of responsibility in order to develop what I will call an integrated theory of responsibility. But this is a constructive argument and not simply a matter of explicating a shared idea of responsibility.

Responsible existence, I argue, aims at respecting and enhancing the integrity of life. The meaning of responsibility depends on the concept of integrity. Derived from the Latin word *integri*, meaning "as whole," the term integrity denotes wholeness or completeness. Integrity also entails a commitment to some project or principle around which the whole of life is seen and evaluated. There can be different kinds of integrity, say personal integrity or professional integrity or even bodily integrity. An athlete, for instance, brings completeness to her or his life through a commitment to physical excellence. The struggle of the moral life is to integrate rightly the complexity of life against disintegration, fragmentation, and also false and tyrannous forms of personal and social unity. Thus, my concern in this book is with *moral integrity*. Moral integrity is defined by an abiding commitment to a specific moral project as well as by the attitudes and dispositions this commitment entails. By this idea I mean the wholeness of life founded on a commitment to respect and enhance diverse goods in personal and social life.[2]

This commitment, I further hold, is the moral meaning of Christian faith. It entails seeing, evaluating, and judging all things in relation to God because the integrity of all existence is known in God alone. This conviction can be formulated as an imperative of responsibility: *In all our actions and relations we are to respect and enhance the integrity of life before God.* Respecting and enhancing the integrity of life is the meaning of responsibility; the integrity of life is the moral good; the source and goal of that good and thus the scope of the moral community is defined in relation to the divine. The basic claim of Christian faith is that in living a life defined by this project, the integrity of one's own life is enacted. By defining our lives wholly in the project of living responsibly before God, that is, "losing ourselves," we paradoxically gain ourselves (cf. Matt. 10:39; Luke 17:33). So defined, a theological ethics of responsibility articulates a distinctive worldview for understanding and orienting life.

The account of responsibility offered in this book insists that moral agents orient their lives with respect to an evaluative framework, and, furthermore, that every evaluative framework entails judgments about the exercise of power. The idea of responsibility helps us to articulate this distinctive moral world-view in which agents exercise power in response to others and with respect to social practices of praise and blame. Responsibility includes beliefs about moral agents, patterns of interaction and responsiveness, and also social practices of praise and blame. The problem with most theories is that they fasten on only one aspect of responsibility. In order to explicate the complex nature of moral responsibility, this book develops an integrated Christian ethics of responsibility consistent with the imperative of responsibility.

It is important at the outset to distinguish an integrated ethics from other forms of responsibility ethics. I want to turn next to the enterprise of ethics itself, since my argument is structured around the dimensions of ethics. In later sections of the chapter we will delineate various theories of responsibility and also basic assumptions in theological ethics. As we will see, this book proposes new directions for Christian ethics in terms of the nature of ethics, the idea of responsibility, and how to understand central claims of Christian faith.

THE DIMENSIONS OF ETHICS

In its broadest sense ethics is the examination of life which seeks to determine how we should live. It is inquiry into character and conduct in so far as these can be considered good and bad, right and wrong. Ethics, then, must define what counts as a moral way of life. To be considered moral a way of life "should include consideration as to what one thinks it important to do and in what ways; how to conduct one's relations with other people; and being aware and prepared to be critical of one's basic approvals and disapprovals."[3] The task of ethics is thus to articulate the values one ought to seek, to assess them critically, and thereby to provide guidance for how to live. An integrated ethics of responsibility defines morality in terms of levels of value, a theocentric principle for choice about actions and relations, and the assessment of identity-conferring commitments. The distinctiveness of Christian ethics consists in its perspective on life, that we should seek the integrity of life *before God*.

Ethics seeks to provide the means to think coherently and comprehensively about how we should live. Method in ethics is how one undertakes the task of systematically coordinating and guiding reflection on the moral life. The organizing perspective of theological ethics, the principle of the integrity of life before God, must be developed, I contend, so that five dimensions of ethical reflection are coherently related.[4] The concept for relating these dimensions of ethics is responsibility. The integrity of life before God is the first principle of theological ethics; the idea of responsibility is central to the method of ethics, that is, the relation between the dimensions of ethical reflection. What then are the dimensions of ethics, and how do they relate to responsibility?

We can clarify the dimensions of ethics with respect to basic questions which constitute the field of morality. The fact that the field of morality is constituted in terms of basic questions also shows us why responsibility is important for the method of ethics. We must respond to these questions in terms of social roles, our existence as agents, and with respect to identity-conferring commitments, the faith we live by. And we answer these moral questions, if at all, not simply in speech, but in actions. As J. R. Lucas writes, "Actions are very largely communications, and carry a meaning which they acquire not by linguistic convention, as with ordinary language, but by virtue of their causal effects, though construed in some conventional social setting."[5] Responsible action is how we communicate the integrity or fragmentation of our lives in response to the questions which life poses to us. This is why the idea of responsibility is basic to the method of ethics.

The basic questions which constitute the field of morality and thus the dimensions of ethics are these: what is going on? what is the norm for how to live? what are we to be and to do? what does it mean to be an agent? and how do we justify moral claims? The questions are basic in the sense that other topics in ethics can – and I think must – be situated among them and must be addressed in a way consistent with how these basic questions are answered. These questions do not rule out other questions. They provoke them. The questions are basic only in the sense that answering them helps to bring coherence to the other things we ask and seek to answer about our lives. Reading the history of theological ethical reflection, in systematic work as well as hortatory literature, reveals that these are questions thinkers ask and seek to answer in examining the complexity of life. Given these basic questions, we can specify interpretative, normative, practical, fundamental, and meta-ethical dimensions of ethics. Let us explore briefly each of the dimensions of ethics. They are not lexically ordered. Moral thinking moves among all the dimensions of ethics.

Any ethic must, first, provide some theory of moral norms, that is, some account of the principles for moral choice and the goods persons ought to seek. This is the domain of what is usually called moral theory; it is the normative dimension of ethics. An integrated theory of responsibility, like any ethical position, provides a theory of value about the good(s) of human life and also a theory of right or norms for moral decisions. I address these matters in Chapter 5. Specifically, we are to respect and enhance the integrity of life before God and this requires attention to the diversity of goods which ought to be integrated in life. In this respect, the relation between the good and the right is the fundamental question which any moral theory tries to answer. I argue that a theory of right is set within a comprehensive theory of value, but, also, that acting on the imperative of responsibility gives rise to a higher, ethical good of moral integrity.

Ethics must, second, examine the nature of human beings as moral agents. Norms and goods are morally irrelevant if agents cannot live by them. The meaning and truth of norms and values must be specified with respect to our being agents in the world with others. We can call this the fundamental dimension of ethics since without agents who act and suffer and wonder about how to live there is simply no point to ethics. The discourse of responsibility, I show in Part II of the book, is uniquely suited for tracing the connection between beliefs about the nature of moral agents and a theory of moral norms and values. Moral agents are those creatures who can and must constitute their identity and thus self-understanding through a set of commitments about how to respect and enhance the integrity of life. This is explained in an integrated theory of responsibility in terms of what I call in Chapter 7 the practice of radical interpretation as the form freedom takes in the moral life. We are free to understand ourselves as responsible agents and thus to have our lives constituted by basic commitments.

Third, any form of moral inquiry must articulate the method and criteria of making practical judgments about what we are to be and to do. In doing so it seeks to show the respect in which moral norms and values bear on the actual lives and decisions of agents and communities. When we say, for instance, that persons ought to be responsible for their actions, the idea of responsibility is meant to convey some claim about how one ought to act and what kind of person one should be. An ethics of responsibility, accordingly, must provide a theory of moral reasoning in order to provide direction for what we ought to be and to do. The concern here is the practical dimension of moral reflection. I address this issue in Chapter 6 in terms of the forms of the assignment of responsibility and in Chapter 7 in a discussion of corporate responsibility.

The moral life is a matter of specific situations and relations in which we wonder about how we ought to live. A moral judgment about what to do or what kind of persons and communities we ought to be is always made with respect to some moral situation. Given this, an ethics, fourth, must develop a theory of how to interpret moral situations and moral discourse. We need to understand what is going on in the moral life. Although I have begun to address these matters in Chapter 1 of this book, I will return to them throughout the text. I will clarify them further by examining the idea of responsibility (Chapter 3) and also theories of responsibility (Chapter 4). In a word, an ethics must aid in rightly understanding our moral situation. The interpretation of a moral situation reflects the normative, fundamental, and practical dimensions of an ethic even as it specifies the context in which particular moral agents are called upon to act. This book shows how the idea of responsibility provides resources for interpreting the deepest impulses and confusions of late-modern societies and the moral meaning of Christian faith.

Finally, ethics asks about the validity of beliefs about the norms and goods of life, moral creatures, practical judgments, and the account of moral situations. Given this, any ethics must provide justificatory arguments for the claims made in each of the other dimensions of ethics. And it must attempt to validate its form of moral reflection and basic concepts. This enterprise is what is usually called meta-ethics. I address these matters in Chapters 8–9 of this book. But throughout the rest of the text, I will show how a conception of human beings as relational agents requires the idea of responsibility, and, what is more, that some idea of the divine is needed in order to understand the sense of responsibility. In this respect, meta-ethical inquiry must be seen not as a prolegomena or addendum to ethics but as an integral dimension of moral inquiry.

This account of the shape of moral thinking should seem obvious simply because it articulates what we already do whenever we think about how to live. It is a picture of moral understanding as the unique way we exist as persons in the world. The reason for this is that the dimensions of ethics aim to answer the basic questions which permeate our existence as moral agents: we seek understanding through interpreting life; human life entails judgments about how to live with regard to orienting goods; we have to decide in particular situations what to do and what kinds of people to be; what is ultimately at issue in our lives as agents is the meaning and value of personal and social existence; and, finally, we ask about the truth of our lives. A vision of human life and its good is refracted through the questions we encounter in relations with others and which we answer with our lives. Ethics provides, then, a moral ontology, that is, an account of the meaning of our being in the world and how to orient ourselves in the world. Theological ethics demonstrates the meaning and truth of Christian convictions in and through the dimensions of ethics in order to provide guidance for how to live. It offers a distinctive theistic moral ontology.[6]

Any ethics will manifest these several dimensions of moral inquiry and will draw on a variety of resources to do so. The methodological problem in ethics is how to articulate the con- nection between these dimensions of moral thinking. What differentiates ethical systems is precisely this point of method. Thus, an ethics of responsibility, I contend, is best defined as that form of moral reflection, which, in distinction to other forms of ethics, uses "responsibility" as the means for relating system- atically the dimensions of moral inquiry. Much as one can speak of virtue ethics, natural law ethics, or divine command ethics, an ethics of responsibility articulates the interrelation of the dimen- sions of ethics in a distinctive way. What is involved in making a case for an ethics of responsibility is then demonstrating that this idea can and ought to serve this function within ethics.

A final point must be made about this account of ethics. The force of this position is to assert that the self-understanding of persons and communities can only be spoken about in and through the interpretative, normative, and practical dimensions of reflection. Persons do not rightly understand the meaning of their own moral existence except by addressing other questions, ones like what ought *we* to do? what is going on in *our* world? and what is the good for *our* lives? We do not understand ourselves and how we should live by simply gazing inward at our lives or by accepting conventional moral beliefs and values. We must examine our lives and come to understand ourselves and how we should live by interpreting our lives and the values we have previously accepted in and through the other dimensions of ethics. The moral point of view means seeing one's existence in the context of this field of moral questions and then critically examining one's life and the life of others from that perspective.

This claim about understanding and the moral point of view has important implications for theological ethics. First, moral understanding is achieved by interpreting and then answering the basic questions specified in the dimensions of ethics. Even distinctively Christian beliefs about the human relation to the divine are only understandable morally when addressed with respect to the dimensions of ethics. This is because Christians claim that the meaning of our existence before God informs how we understand moral situations, specifies the norms and values for how to live, and bears on specific moral judgments. The dimensions of ethics structure moral thinking whether or not one is interested in the Christian vision of life. And, since moral understanding is specified through the dimensions of ethics, the theologian must show the meaning and truth of faith with respect to these very same dimensions of reflection.

The second implication of this account of the nature of moral understanding concerns the unique good of moral integrity. If we come to some critical self-understanding by examining how and why we answer these morally basic questions, then the form of identity achieved through this process of interpretation is yet another level of human existence. Through the activity of radical interpretation, as I call it in Chapter 7, we determine the desires and volitions we want to characterize our lives. An integrated ethics of responsibility thus isolates two distinct but related sets of goods. First, there are goods, which I explore in Chapter 5, entailed in pre-moral, social, and reflective levels of life. But, second, there is the good of the moral life itself in so far as it is constituted through a commitment to respect and enhance first-order goods. The idea of moral integrity developed at length in this book denotes this higher good of human life. The responsible life aims to respect and enhance the integrity of life and in doing so enact the good of moral integrity.

The third implication of this account of moral understanding is that while moral claims are unavoidably self-referential they are not simply expressions of personal preferences or social conventions. Moral values, interpretations of moral situations, and decisions about what to do always refer to the agent who is evaluating, understanding, and acting. However, since the self-understanding of any particular moral agent, or community of agents, is constituted with respect to the questions articulated in the dimensions of ethics, the meaning and status of answers to those questions cannot be simply dependent on the self and its preferences. Moral understanding has a realistic intention: we seek to understand what is going on in a moral situation; we want to assess and comprehend the values or norms which have informed our self-understanding; we need to know what to do in specific, concrete situations of choice. Moral claims refer to some agent or community because these claims are only intelligible when apprehended through the medium of our language, traditions, beliefs, and forms of life. But interpreting these claims shows, or seeks to show, that valid moral norms, accounts of moral situations, and decisions about what to do are not reducible to the subjective understanding of an agent. I have called this position hermeneutical realism in ethics.

This account of the shape of ethics and the implications it entails for moral understanding, the moral good, and the status of moral claims are presented in the following chapters. What we see now is that the idea of responsibility is basic to method in ethics because it articulates the field of morality through distinct, but interrelated, dimensions of reflection. The point for theological ethics is to insist that Christian claims must also be interpreted and assessed with respect to the very same dimensions of ethics.

THEORIES OF RESPONSIBILITY

All previous theories of moral responsibility can be grouped into three types: agential, social, and dialogical. First, there are theories, the agential, which ground responsibility in the acting agent. This type of theory focuses on the connection between an agent as a causal force in the world and the evaluation of her or his acts. An agent is accountable for the deeds she or he does. I am responsible for the promises I make, for instance. The central ethical conundrum is then the problem of freedom and determinism with respect to deeds for which agents can be rightly held accountable. The second type of theory, the social, centers on social practices of praise and blame. Responsibility is grounded not in a causal relation between an agent and her or his deeds, but with respect to the roles and social vocations an agent holds. The president of a corporation, for instance, can be responsible for actions he or she did not directly do. Given this, the moral community and the boundaries of that community become the focus of moral attention. The third type of theory, the dialogical, focuses on the event of encounter with others and thus on that to which or to whom one is responding. I am constituted as a moral being in terms of how I respond to the claim of others on me. For this type of theory, human life is primarily a matter of responsiveness to others. Thus, agential theories of responsibility center on the agent/act relation, social theories on social practices, and dialogical theories on the self/other encounter. At best there are family resemblances among these theories of responsibility since they are clearly not examining the same basic phenomenon or experience.

One task of this book is to explore questions of responsibility with respect to these types of theories of responsibility. My contention is that each of these theories articulates a genuine insight into the moral life, but that none of them is adequate for addressing the theme of responsibility. Theories of responsibility which center on the acting individual often fail to address the importance of social roles in assigning responsibility due to an overriding concern to protect personal autonomy. Theories which concentrate on responsibility in social roles often fail to explore the inwardness of our lives as individuals. And an ethics of responsibility which focuses moral attention on interpersonal encounters too easily reduces questions of responsibility to personal responsiveness. Obviously, we are agents who act and exist in social relations, and we are also called to respond to others, even the divine other. The difficulty is how to account for each of these facts of life within an ethics of responsibility. We need an ethics which can draw on the insights of these types of theories without reducing its scope to any single aspect of responsibility. And this is what I mean by an integrated theory of responsibility.

The focus of an integrated theory of responsibility is the connection between commitments basic to the identity of agents, or communities of agents, and the power to act in relation to others and the environment. The norms for the use of power articulate projects to which responsible agents and communities are dedicated as basic to the meaning and coherence of their lives. The work of ethics centers then not simply on the relation of an agent to her or his acts, the character of social practices of praise and blame, or the event of encounter. Rather, ethics addresses the question of the project to which we ought to commit our lives. For theological ethics this is the question of the moral meaning of Christian faith.

Plainly, there are divergent theories of responsibility. Yet it is also true that the idea of responsibility can play different roles in ethics. This leads to different paradigms in the ethics of responsibility. By paradigm I mean a coherent and comprehensive way of examining some subject-matter, in this case the moral life, with respect to a root metaphor or first principle which serves to organize the inquiry. The first principle, or root metaphor, of an ethics is the primary and irreducible source for the intelligibility of the moral life and moral values and norms. It is the principle of coherence for thinking about how we can and ought to live; it is the conceptual and thematic heart of an ethics.

While questions about human responsibility will be found in all moral reflection, that does not mean the idea of responsibility must function as the first principle of an ethics. An ethics might specify the idea of utility, happiness, the Good, duty, or some other central norm, and then treat the questions of responsibility within that framework. In these cases, an ethics of responsibility would mean simply an examination of the conditions, scope, and legitimacy of moral accountability. But the norm of moral accountability and the paradigm for ethical thinking would be established on grounds other than the idea of responsibility. A good deal of work in ethics on moral responsibility is undertaken in this way. In such work, responsibility does not define the moral point of view; it is treated within the context of a normative framework specified around another first principle.

However, an ethics of responsibility can also mean that responsibility is the "basic, irreducible conception which serves as a starting point for the development of a coherent and comprehensive ethical doctrine."[7] Responsibility in this case is the first principle for an ethics. To be sure, this form of ethics will treat particular questions of moral accountability, but they will be addressed from a point of view defined with respect to the idea of responsibility. For this form of ethics, the central question becomes the meaning of responsibility itself. As the first principle of the ethic, the idea of responsibility specifies in shorthand terms a moral outlook and attitude. I call this a "strong" as opposed to a "weak" ethical paradigm in which the idea of responsibility is not the first principle of ethics.

Ethics of responsibility can be classified, then, in terms of moral focus (agential, social, dialogical) and with respect to the first principle of an ethics (strong, weak). These distinctions help us to understand work within Christian ethics on the theme of moral responsibility. Theologians earlier in this century like Karl Barth, H. Richard Niebuhr, Bernard Häring, and others developed dialogical ethics of responsibility. Reflection on moral responsibility was dominated by a picture of the moral life in terms of responsiveness to others. To be responsible did not mean simply to be accountable for actions, but, rather, to respond rightly to others and also obediently to God. Richard Gula, a Roman Catholic ethicist, locates the dialogical pattern in Jesus' message itself.

Jesus's message is an announcement calling for a response. The whole moral life is a response to a call, to the divine initiative. Call–response forms the structure of the moral life in the message of Jesus, though nowhere does Jesus, or the New Testament at large, provide a moral system as such.[8]

Persons are called to respond to God's action in the world and in Jesus Christ. The moral problem from the Christian perspective is how to respond rightly to others in light of God's action. The task of the theological ethicist, accordingly, was to develop a moral system consistent with the call–response structure of the moral life as presented in the message of Jesus. Theologians, as we see in Chapter 4, developed strong and weak versions of a dialogical ethics of responsibility. The pattern of call–response was the basic paradigm of thought for virtually all Christian ethics of responsibility.

This book presents an ethics which differs from dialogical theories in how responsibility is understood, and, furthermore, with respect to what is properly the first principle of ethics. In terms of the method of ethics, the idea of responsibility provides the means for thinking coherently and comprehensively about the whole of the moral life. However, responsibility does not specify the moral good for the ethics proposed in the book. The problem of personal and social integrity, that is, the integration of complex elements of life with respect to a commitment about the worth of finite life in the exercise of power, is the focus of moral attention. Moral integrity is the substantive moral good and hence focus in theological ethics; the idea of responsibility provides the means for thinking about the meaning of that good for how we ought to live. Neither a "strong" nor "weak" theory of responsibility, the position of this book is what I have called an *integrated* theory of responsibility.

This distinction between method and content in theological ethics further distinguishes my argument from previous accounts of responsibility in Christian ethics. Theologians earlier in the century took the idea of responsibility to denote not only how one ought to live, the content of morality, but also how to think about the moral life from a theological perspective. Ethical thinking, the argument went, is dependent upon our encounter with what is other than our thinking and our existence. This made revelation, the action of the other encountering the self, primary in moral thinking. Given this approach to ethics, all moral concepts which did not fit within a call–response model or could not be reformulated in terms of that model were seen as beyond the scope of theological ethics. Not surprisingly, concepts like virtue, happiness, and moral perfection, to name a few, simply disappeared from Christian moral discourse. The dialogical paradigm was unable to deal with the full complexity of our available moral discourse.

The dialogical paradigm must also be reconsidered because it is unable to address the complexity of life in a late-modern, technological world. As we noted in Chapter 1, the radical extension of human power coupled with the reality of moral pluralism is what poses the need for a new ethics of responsibility. We are not simply responding to others, even a divine other; we must determine the goods we ought to seek through the exercise of this power and the norms which ought to govern its exercise. The social world is also increasingly differentiated into distinct but interacting subsystems, for instance, the economic, media, legal, and political systems, and it is riddled with the problems and possibilities of moral and religious diversity. A dialogical model of responsibility based on the I–Thou encounter simply lacks the resources to address the complexity of the late-modern social world. It reduces the central moral datum to interpersonal relations and thereby fails to grasp how the exercise of power, the meaning of moral norms and values, and even the capacity of persons to act are increasingly mediated by the social world. The central moral problem in our context is how to exercise and value the power to act in diverse social spheres and institutions without making power and the will-to-power the standard of the good.

To act responsibly, I argue, is not simply to respond rightly to the "other." It is to put power in the service of a meaningful life rather than to have the meaning of personal and social existence defined by the exercise of power. The central conundrum for ethics, accordingly, is how to show that at one and the same time the power to act is integral to meaningful human life, and yet that this good is subject to moral evaluation and not the good itself. This is an ancient problem in Western ethics. In the *Gorgias*, Socrates confronted the claim of Callicles that the powerful can, must, and, in fact, do define what is morally right and good. The Socratic enterprise was to counter this thesis with the claim that moral goodness is not reducible to the will-to-power. In the biblical world, the prophets challenged cultic meaning-systems which seemed impervious to the oppression of the poor and the weak. True faith, the prophets insisted, is seeking justice, loving mercy, and walking humbly with God (Micah 6:8). In modern thought, Max Weber, in his famous essay "Politik als Beruf," formulated an ethics of responsibility with respect to the demands of political leadership.[9] Weber contrasted an ethics of absolute conviction with an ethics of responsibility. For Weber responsibility meant, as Wolfgang Huber notes, "the liability for future effects of present actions" given the unique power of political leaders.[10] A host of current thinkers are renewing Weber's concern for power and the future effects of actions. The debate about the relation between power and moral goodness cuts like a knife through the history of Western ethics.

The conundrum about the value of, and norms for, power is answered in Christian ethics by means of two interlocking beliefs. First, the God of Christian faith is ultimate power which permeates and actuates all that is, and yet this power is identified with respect to the affirmation of finite reality in the midst of history. To identify or name ultimate reality "God" is, I argue, nothing less than a radical interpretation of power which asserts that power does not define the good.[11] The theological ethical task of naming God, that is, thinking the reality of God through the diverse literary and symbolic forms of religious discourse which render the divine identity, is to articulate how Christian beliefs about "who" God is transvaluate power. The theologian then seeks to show the ways in which this construal of reality can and must transform the lives of agents. In theological ethics this means that the good is defined by a relation between power and finite being, a relation in which finite existence is respected and enhanced in its finitude.[12] "And God saw that it was good," as we read in the creation narrative (cf. Gen. 1:1–5). Faith in God organizes the whole of Christian life and entails a radical transvaluation of the value of power. This means, second, that Christians are committed to seeing the creator in the glass of all creatures and thus to respect and enhance the worth of all finite reality. This form of moral understanding, I want to show, entails a perception of value and also a value judgment of the most radical kind. It is to see and judge the worth of finite beings in relation to the divine identity and then to undertake a life project consistent with that insight and judgment.

Christian faith provides an evaluation of power. It also warrants a specific construal of the world. In this way, the ethics of responsibility provides conceptual and evaluative resources for the interpretation of our social and moral context.

BASIC ASSUMPTIONS

A treatise in theological ethics rests upon assumptions about human existence, the meaning of religion, and resources essential for moral and theological reflection. This book draws on a variety of resources in ethics, including scripture, Christian tradition, work in philosophical and theological ethics, and also social and political theory. What must be clarified now are the basic assumptions about human life and religious beliefs which orient the argument. At the root of these assumptions is the conviction that we need not derive all of our knowledge of human existence from Christian thought alone. What is more, for the purposes of ethics, Christian faith must be understood historically and comparatively with respect to other religions and belief systems in the world.

The basic assumption about human beings at the heart of this book is simple enough and hardly original. As persons, we understand ourselves as agents in relation to others and the world, and, further, our lives manifest a relentless quest for wholeness. We encounter the fragmentariness of existence in two interrelated ways: first, we are parts of a larger reality on which we depend and against which life contends and also adapts; and, second, in the depths of our lives, in the chaos and complexity of personality, we are also fragmented yet seeking wholeness. It is the Christian conviction, as St. Augustine put it, that the restless quest for harmony with the whole of which we are part and for peace within ourselves is the inner meaning of human existence. The quest finds its true end in the divine. Yet even if one rejects that conviction, the fact of human aspiration, as well as its denial and distortion, cannot be denied. Human aspirations are the creative impulse in personal and social life and also the deep roots of human conflict. Human beings seek wholeness through the exercise of power. Personal and social life manifest the quest for access to value creating power – the power to integrate existence and thus fulfill the restless quest of existence.

This fact is also basic to an understanding of religion. If we look at religions historically and descriptively rather than dogmatically, it seems clear that the idea of power or powers is central. Any form of religious conviction entails beliefs held by persons and communities about "power or powers operative in their own lives and, they presume, also in nature beyond them."[13] The religious sensibility is characterized by a sense that there are powers which create possibilities for life and with which human beings contend. Religious communities are characterized by ritual means for acquiring and releasing power. And the religious life is an openness to personal, even mystical, experiences of empowerment and radical transformation at the center of human existence. This means that religious activity is always a response to what is experienced as an active force in the world and in human existence. To be sure, religious traditions develop complex systems of ideas, beliefs, symbols, and narratives to speak about the "object" of experience. But these too can be understood with respect to the religious outlook and sensibility.

It is also the case that how any particular historical epoch understands the human relation to powers other than itself will differ. A crucial point of difference between classical and contemporary modes of thought is precisely on this point. In the ancient world the "powers" to which human beings were responding were understood in agential terms. The universe had to be understood with respect not only to human agents, but also to supra-human agencies. The idea of purpose was then basic to the classical worldview. This conviction has been rejected by the modern world. Modernity has insisted that there is but one kind of agent in reality, the human agent, and the world is the place of this agent. The human is on its own, this life is all there is, and human beings must assume responsibility for their lives and the lives of humankind. Classical religious and metaphysical accounts of reality no longer seem to help us explain the way the world goes, but actually hinder our understanding of the processes of reality.[14] Modern science has rejected some final purpose, a *telos*, as crucial for explaining reality.

Theologians earlier in this century attempted to respond to this critique of mythico-agential and teleological accounts of reality in order to speak of God's action in history. One of the ways they did this was through some version of the dialogical model of responsibility. The Word of God encounters the I as a Thou in a free event of address (Barth), or God is the "third" through which the self–other relation is to be understood (H. R. Niebuhr). It is, then, in the word–event, the linguistic announcement of the Word of God, that God "acts," or, conversely, we are to understand the divine with respect to our social, dialogical interactions with others. By dropping mythico-agential accounts of the natural world, these theologians attempted nevertheless to make sense of traditional Christian claims about God acting in history. In so far as this book provides a new ethics of responsibility, it should be clear that claims about the divine will be interpreted in terms other than the I–Thou encounter. The point of continuity with the work of these previous theologians is the conviction that an ethics of responsibility is the most adequate means for understanding the human relation to the divine amid the moral life.

I argue that the theological ethical task is to interpret basic Christian ideas and symbols in order thereby to understand and guide human life in the world. Since an ethics of responsibility focuses on the exercise of power by agents and communities in responding to others and the world, the question then becomes the meaning of theological claims for matters of power and responsibility. Agential accounts of the divine are important in this respect not because they literally designate a supreme agent, but because to speak of ultimate power in terms of agency is to assert a value beyond power. The most radical claim of Christian faith, I hold, is that ultimate power, the divine reality which creates, sustains, judges and redeems all of existence, is known and understood as binding its identity to the respect and enhancement of finite reality. Christian claims about "God" are interpretations of power which render all exercises of power subject to moral evaluation. One can thus speak about a "personal" God in order to specify the belief that power alone is not the ultimate value, but not to give causal explanations of the physical world. Once this point is fully understood, then the ethicist can specify the moral meaning of theistic discourse without reverting to mythico-agential accounts of reality. Understanding human life and the world from this point of view entails seeing and evaluating all things in relation to a creative power which pervades and actuates reality and which binds its identity to the integrity of life. To borrow a phrase from Paul Ricoeur, the symbol "God" gives rise to moral thought about the transvaluation of power.[15] It is this insight, I contend, that was not thoroughly grasped in previous theories of responsibility in Christian ethics that centered on the event of encounter. Yet the insight is essential for an adequate ethics of responsibility in our time.

The theory of responsibility developed in this book aims then at securing the value of finite existence through the exercise of power to respect and enhance the integrity of life. Human beings seek and struggle for some measure of integrity or wholeness in their lives and their communities; the ethical question is how to specify the integrity of life which ought to be sought in all our actions and the norms which should govern our conduct in that quest. The idea of responsibility, I hope to show, requires the transformation of what persons and communities seek in life, their vision of the good, and the projects to which they are committed. Responsibility is the concept we can and must use to sort out the complex relation of power and value in the domain of individual and social action.

These claims about human existence, the religious mentality, and also matters of responsibility bring us to another basic assumption. While the increase of human power makes responsibility central to current ethics, power is not the subject-matter of ethics. The ways in which human beings and communities ought to live in relation to the God of Christian faith is the subject-matter of theological ethics. In this respect, Christian ethics resists any denial of the dignity of persons in thinking about life and the world. Yet for some contemporary theorists reflection on power virtually eclipses concern for inquiry into human existence. We are told, for instance, that forms of human subjectivity are the product of the networks of power operative in a society, and, given this, those power networks, and not human existence, are the proper topics of reflection.[16] This means that the assessment and direction of power is a social and political question, but not fundamentally a moral one. Within the enterprise of developing a theory of responsibility another feature of this book is an abiding concern for human agency and moral identity even though the problem of power in its various forms is indeed central to any plausible ethics of responsibility in our time.

I will return later in Chapters 6–7 to the question of how to understand moral identity. Theological ethics, I show, aids in transforming and reconstituting how we understand and intend our lives as moral beings in the world with others. We human beings are creative interpreters of ourselves, the world, and our lives together. We do not simply conform ourselves to a pre-determined script etched in human "nature" which ought to be lived out come what may; we are also not infinitely plastic beings who can be fashioned and remade at will by networks of social and cultural power. Human beings are defined as agents intimately related to each other and to their environment seeking integrity in life. This makes responsibility basic to human existence.

By examining commonly held beliefs about responsibility, the discourse of responsibility, and also various theories of responsibility I will attempt to isolate and articulate the meaning of responsibility. Moral responsibility, I argue, specifies the relation between moral norms and the identity of some agent exercising power in the world. The exercise of power by persons and institutions is morally responsible only when it respects and enhances the integrity of life. Responsibility is bound up with attitudes and projects to which individuals and communities are committed as essential to the meaning and value of their lives. The question of the moral life is accordingly about the project to which we ought to dedicate ourselves and how that commitment constitutes our identity and an understanding of the world in which we live.

CONCLUSION

In this book, I develop an account of moral responsibility from a Christian theological point of view by engaging other positions around central questions in ethics. The position I forward cannot be fairly described as focusing on the acting agent, on the demands of responsibility which surround social roles and practices, or on the I–Thou encounter. These are the usual options in the ethics of responsibility. All of these theories make contributions to an adequate theological ethics of responsibility, and, yet, none of them alone can address the complex set of problems and possibilities which surround responsible human life. I have tried to incorporate aspects of each of these theories into an integrated theological ethics of responsibility which can clarify the moral meaning of Christian faith, a theme in this chapter, and respond to the pervasive moral problems of our time explored in the previous chapter.

II

The theory of responsibility

CHAPTER 3

The idea of responsibility

Part I described the contemporary moral landscape and showed why an integrated theory of responsibility is needed in ethics. This theory must account for the deep concern for fulfillment and authenticity in contemporary culture in such a way as to make sense of beliefs about the world and human life captured in the idea of responsibility. Part II of the book presents an integrated theory of responsibility. As the first step in developing such a theory, I want to examine in this chapter the semantic and conceptual field demarcated by the idea of responsibility.

THE LANGUAGE OF RESPONSIBILITY

The importance of responsibility to moral reflection is a relatively new development in Western thought. In the modern sense of the term, the word responsibility is not basic to biblical morality or Greek and Roman ethics. The term responsibility does derive from Latin, however. The Latin root signals the diverse meanings of responsibility. *Respondeo* means to promise a thing in return for something else, and also, in legal discourse, to give an opinion, advice, decision, or, generally, an answer when one is summoned to appear in court. This meaning of responsibility is reflected in the German *Verantwortung*, where the emphasis falls on answering (*Antwort*). A responsible agent is one who can answer (be responsible) for his or her actions and intentions before someone who questions the agent, even if that "someone" is the agent himself or herself. In answering the self appears and is present before whomever she or he must respond. In answering, the bond between person and her or his deeds is acknowledged. Histori-

55

cally this simply meant that a person had to appear before some
legal, moral, or ecclesial tribunal.

With some reflection we can grasp a more profound meaning
of this sense of responsibility. The "self" appears in the activity of
answering and responding to others. The activity of answering is
then the primal deed in which the moral self is enacted. This led
H. Richard Niebuhr to speak of human beings as dialogical
creatures, *homo dialogicus.*

> In trying to understand ourselves in our wholeness we use the image of
> part of our activity; only now we think of all our actions as having the
> pattern of what we do when we answer another who addresses us. To
> be engaged in dialogue, to answer questions addressed to us, to defend
> ourselves against attacks, to reply to injunctions, to meet challenges –
> this is common experience.[1]

The self exists in answering and responding to others. In this
sense, others are a necessary condition of being a self. Responsibility touches the whole of life since what it means to be a self is
enacted in responding to others.

While many theories of responsibility concentrate on the
activity of "answering," the other sense of the Latin *respondeo* is
also important. The idea of responsibility has been set within debt
language, what one owes another in legal and economic matters,
and thus the debt of justice. Here the emphasis is upon obligations which arise in promissory contexts. The agent must give an
account of actions with respect to some promise or debt made to
others or to himself or herself. Accountability can be seen as a
specific instance of the activity of answering; it is a particular form
of answering for oneself to others which falls within the rubrics of
debt and duties. This is why responsibility is often taken as
synonymous with being accountable for something or some
action with respect to what is justly owed to others. Accountability
is also important for the discourse of praise and blame; we
normally do not praise or blame persons who are not accountable
for some act or event.

Responsibility has its roots in legal, economic, political, and
personal contexts. It also has roots in the arts. There is an early use
of the word in English for an actor who is willing and able to play

any part required by the company. Such a person was called a "good all-round responsible." This theatrical use of the term, along with the idea of "character," designates the importance of social roles in assigning responsibility and also shows that we can bear responsibility for others, act in their place and for them. To speak of a "good all-round responsible" focuses attention on the role(s) an agent must play in some social activity and also that persons can assume the roles, and thus responsibility, of others. In a sense the self simply *is* the roles he or she plays and performs, roles which entail complex patterns of relations to others. Max Weber, as I noted before, drew on this insight for an ethics of responsibility in political life. The politician, by virtue of his or her office, must make decisions not simply in terms of absolute moral norms, but with respect to the future effects of present action.

The meaning of responsibility as representative action, acting for others, was especially emphasized in Christian ethics by Dietrich Bonhoeffer. In his *Ethics*, Bonhoeffer notes that

[the] fact that responsibility is fundamentally a matter of deputyship is demonstrated most clearly in those circumstances in which a man is directly obliged to act in the place of other men, for example as a father, as a statesman or as teacher.[2]

Bonhoeffer understood the idea of responsibility in terms of Christ's action, that is, Christ is the one who has acted in our stead. The Christian is to act likewise. The Church also is to undertake this moral task with respect to the wider culture. Yet Bonhoeffer's point holds even if we bracket his ecclesial and Christological claims. In many situations responsibility requires that someone act in the place and on behalf of another. In terms of our offices or vocations, we are often called to be "good all-round responsibles" with respect to the values and obligations entailed in our social roles. We speak not only of the responsible self who answers for its actions, but also the responsible official, teacher, priest, parent. As Dorothy Emmet notes, a "role is a capacity in which someone acts in relation to others. It is, of course, a metaphor from the theater, where a role is a part assumed by one actor in a play where others assume other parts." But what "people think they ought to do depends largely on how they see their roles, and (most importantly) the conflicts between their roles."[3] Later we will see that this is the central claim of social theories of responsibility.

The semantic field of responsibility includes representative action, answerability, and also accountability for actions. I will return to these meanings in more detail below. Yet in spite of its Latin root and diverse meanings, the word responsibility is a late arrival in Western ethics. It first appeared in German, English, and French in the seventeenth century. As Albert Jonsen has noted:

> The word has its philosophical debut in David Hume's *Treatise of Human Nature* (1740): "actions may be blameable ... but the person not responsible for them." It is from the beginning used in political literature, as exemplified in Alexander Hamilton's *Federalist Papers* (1787): "Responsibility in order to be reasonable must be limited to objects within the power of the responsible body."[4]

From the outset the idea of responsibility was linked in ethics to persons and actions as well as to the problem of specifying the limits of accountability with respect to the power of an agent to act.

Because of its late arrival in the moral lexicon of the West, the idea of responsibility manifests the complex, social process of modern history. Jacques Henriot observes that responsibility is, first, part of the humanization of Western moral discourse in the sense that only human beings are subject to having their actions imputed to them and thus held accountable.[5] As we saw in Part I of this book, this has made it difficult to speak of the "agency" of God in modern thought. The modern idea of responsibility entails a naturalistic view of reality in which events are explained in causal terms and in which human beings appear as unique creatures because of our capacities for action.

Second, the language of responsibility has contributed to the growth of individualization in Western societies. In ancient cultures it was possible to speak of the community bearing responsibility for the deeds of individuals. For instance, in the Bible we find the idea of the scapegoat who is driven into the wilderness bearing the sins of the people. We read that the sins of the fathers are visited on the children to the seventh generation. The idea of collective responsibility is also seen in the Hebrew prophets. In Isaiah 43:10 the people of Israel are addressed as a single agent. "You are my witnesses, and my servant whom I have chosen." Likewise, in Hosea Israel is compared to a harlot. In so far as Israel has gone after other gods, played the harlot and forsaken Yahweh, responsibility is imputed to the community. The sacrificial, cultic practices found in ancient cultures rely on the concept of corporate responsibility and the shifting of responsibility to a scapegoat. These social practices presuppose that one can assign responsibility to a human collective.

Democratic polity and liberal thought can be defined in relation to a shift from the community as the center of moral attention to that of the individual. The idea is that the individual is a "self-governing agent" and that morality is a relation between social mores and personal freedom. This shift to the individual brought with it, third, an interiorization of the moral life. A person is responsible for himself or herself before the demands of conscience or his or her life-plans. Each individual must decide what values to endorse and what kind of life to live freed from the demands of communal *mores*. This has meant that much modern ethics understood persons as bearing rights rather than being defined by roles and relations. The language of responsibility has contributed to modern thought in terms of specifying actions for which a person can and cannot be held rightly accountable. It has led, Henriot argues, to the differentiation of legal and moral claims. A person should not be held legally responsible for her or his moral failures, although this belief is a relative newcomer in Western thought and life.

Despite its presence in the historical development of the modern world, the idea of responsibility has been given different and even conflicting expressions in ethics over the course of the last two centuries. These manifest deep tensions in modern Western societies. Interestingly, the differences indicated in the meanings we have found in the Latin *respondeo*. The central difference in the early use of responsibility in ethics was between positions focused on personal agency and those centered on responsibility in social roles.

F. H. Bradley, in a famous essay "The Vulgar Notion of Responsibility and its Connection with the Theories of Free Will and Determinism" (1876), used the idea synonymously with accountability, liability, and imputability.[6] Bradley's ethics centers on the idea of self-realization. Yet he holds that one's station or role in social life determines the obligations and values one *ought* to abide by and seek. For Bradley, the task of the moral life is to internalize the values and duties of one's station; it is to know one's station and its duties. The purpose of the moral life is self-realization, but this must pass through the social context of human existence. In the development of the discourse about responsibility within ethics, Bradley examined the connection between responsibility and accountability that is reflected in the Latin *respondeo*. He acknowledged the importance of freedom and accountability for actions, since these ideas are rooted in widely held moral beliefs. But in explicating the meaning of the moral life, Bradley stressed one's social role or station. He outlined the first social theory of responsibility.

At roughly the time Bradley was writing, L. Lévy-Bruhl, in *L'Idée de Responsibilité* (1844), explored the idea in moral and legal discourse. Lévy-Bruhl argued that "the voluntary acts of human beings are imputable to them, and the imputability of actions corresponds to the responsibility of the one who acted."[7] In moral discourse, as opposed to legal thought, responsibility, Lévy-Bruhl insisted, is linked to the idea of conscience. This connection was crucial for specifying the distinctive nature of moral agents. Drawing on Kantian ethics, Lévy-Bruhl's point was that in conscience an agent holds himself or herself responsible for actions. The moral self appears before the tribunal of the conscience; the roots of moral obligation are then internalized. Thus Lévy-Bruhl, in contrast to Bradley, stressed the individual and the demands of conscience even though both of these thinkers examined the relation between responsibility, accountability, and obligation.

In terms of these early developments of the idea of responsibility we can see its central meaning in a standard philosophical dictionary.

Responsibility means (1) the moral obligation, sometimes sanctioned by law, to repair the harm done to another; (2) the situation of a conscious agent with regard to those actions which he has really willed to perform. It consists in his being able to offer motives for these acts to any reasonable person and in his being obligated to incur praise or blame for them according to the nature and value of these motives ... It is the solidarity of the human person with his actions, the prior condition of all real and juridical obligation.[8]

What is at stake in thinking about responsibility is how to understand moral freedom, the solidarity of an agent with his or her acts, and also practices of praise and blame. This is why the discourse of responsibility has contributed to the processes of humanization, individualization, and internalization in modern Western life. The point of difference in early moral theories was whether individual conscience or social role was central for addressing matters of responsibility. This seems to express a deep rift in modern Western societies about how to understand the identity of persons and the values and norms which ought to guide their lives. It brings us to developments within ethics in the twentieth century.

The idea of responsibility has been given conflicting interpretations in twentieth-century ethics. The problem is that the idea of a person is ambiguous and has been given different interpretations, which, in turn, are reflected in theories of responsibility. If there is general agreement that an agent is someone who makes decisions and acts, what specifies individual identity remains in dispute. For some thinkers, identity is defined by the subjectivity of the person. Being a self entails the capacity to use certain pronouns reflexively in response to the question "who is acting?" In answering that question, the self appears through the use of specific linguistic forms. For instance, if I utter the sentence "I broke the promise," then the connection between the reflective first-person pronoun "I" and the particular deed is manifest in such a way as to specify me as a person: I am the one who broke his promise. My identity appears, as it were, in uttering this sentence. I exist in the event of discourse through answering others. The importance of this form of assigning responsibility will be explored in later chapters of this book.

However, it should also be noted that "person" derives from the Latin *persona* meaning an actor's mask or a person in a play. Thus other thinkers, beginning with Bradley, have explored the nature of personal identity with respect to the roles individuals play in the social order. Here the capacity to refer to the self is not dependent simply on the individual as an origin of decision making and action and the ability to answer the question "who is acting?" One becomes a self in and through the appropriation of a social role along with the duties, virtues, and values that role entails. I am a father. I am a member of the United Methodist Church. I am a university professor. Demographically, I am a white middle-class male. Each of these roles and social stations serves to constitute my identity and sense of self. As traditional Protestant ethics put it, persons have various callings and vocations. A person's calling provides guidance for what she or he ought to do. I should fulfill my calling as a professor, for instance. But the calling also shapes character. We identify ourselves and others with respect to the roles we adopt and inhabit.

These problems about how to understand the idea of a person, and thus what appears to be the condition for all discourse about responsibility, have been debated throughout the history of the idea of responsibility. I will return to them in Chapter 7. Yet in this century the processes of humanization, individualization, and interiorization intertwined with the early development of the language of responsibility have been challenged with respect to beliefs about persons. In the face of massive social conflict, political heteronomy and tyranny, and the increasing sense of the social isolation and anomie of modern individuals, the question of the meaning of human existence was posed with grave force by thinkers early in the century. The idea of responsibility remained central to this project, even if there were fundamental shifts in how to conceive of it.

Drawing on the work of the Jewish thinker Martin Buber, many theologians defined responsibility in terms of how a self responds to or answers a "Thou." Responsibility depends not on the solidarity of the person with her or his actions or social roles, nor on the reflexive use of pronouns in answering for one's conduct, but on the intrinsic relation of I and Thou. Buber insisted that human beings utter but two basic word-pairs: I–It and I–Thou.[9] The "I," the self, is constituted by being addressed by a "Thou." Christian theologians adopted this insight. As Emil Brunner put it:

> The being of man as a person depends not on his thought but on his responsibility, upon the fact that a supreme Self calls him and communicates Himself to him … Only in this double relation, not in his rationality, has man as a person his origin and fundamental being. His deepest nature consists in this "answerability," i.e., in this existence in the Word of the Creator.[10]

Theologians insisted that the message of Jesus was an announcement calling for a response to God. The shortest expression of the "call–response" structure as definitive of the meaning of the life of faith is the summons "Follow me!" uttered by Jesus to his disciples (cf. Mark 1:17–18; Matt. 4:18–20). The life of faith is defined by one's response in total obedience to that summons.

While we will explore this understanding of responsibility later, the point is that persons are constituted as selves not through intentional acts or rational self-descriptions, but within dialogical relations and through representative actions. The self "appears" in answering another and acting for others; we become selves in patterns of interaction and our actions are genuinely moral when we act with and for others. As Amitai Etzioni has recently noted, "if individuals were actually without community they would have few of the attributes commonly associated with the notion of an individual person."[11] The self needs a Thou in order to be. This dialogical model of responsibility attempted to circumvent the reduction of the moral life to one's role or social station or to the autonomy of the self.

The dialogical model of responsibility is not the only way persons have been understood in twentieth-century political and moral thought. By concentrating on the "solidarity of the human person with his actions" liberal theory and also existentialist philosophy defined responsibility in individualistic terms. This continued the processes of interiorization, humanization, and individualization in modern Western societies. On this reading, responsibility expresses the modern stress on personal identity and self-consciousness as well as the insistence on moral autonomy. Moral autonomy is understood both in the sense of persons having a right to be free from tyranny and in the sense of the self as the origin of actions. These theories of responsibility emphasized the capacity of human beings to intervene and to change the natural and social world for their own purposes.

An extreme form of this line of thought is seen in the work of the existentialist philosopher Jean-Paul Sartre. He presented a picture of human life as self-creating. In *Being and Nothingness* Sartre argued that

> man being condemned to be free carries the weight of the whole world on his shoulders; he is responsible for the world and for himself as a way of being. We are taking the word "responsibility" in its ordinary sense as "consciousness (of) being the incontestable author of an event or of an object."[12]

In every decision to act the self bears the burden of the life it chooses, whether that life is authentic or inauthentic. In fact, the self *is* this act of decision. The self is thus radically free to constitute itself anew in each and every decision to act. The central moral problem is simply to choose to be a self. As we will see in Chapters 6–7, critics of this rather exalted notion of freedom rightly argue that it places undue burden on agents, and, furthermore, is actually not a crucial condition of moral identity. But Sartre's point was that the idea of responsibility is meant to articulate the consciousness of an agent being the cause of some action, event, or state of affairs. In this respect he simply repeated common usage about the idea of responsibility. Sartre's radical claim was in the implication he drew from this usage. The self does not exist outside of its actions; the self is its act. Existence precedes essence, as Sartre and other existentialists insisted.

This existentialist interpretation of responsibility requires that we isolate the integral connection between self-consciousness and the capacity of an agent to be a causal force in the world. Human beings are no deeper or more stable than their actions. Yet in so far as only human beings can meet the demands of action, persons, if they are to live authentic lives, must recognize that they bear the weight of the world. This is consonant with the process of humanization we have been tracing, a point Sartre actually endorsed. It is also consistent with the fact that the modern world is a disenchanted one. The person is self-creating and thrown in an otherwise meaningless world, a claim the dialogical understanding of responsibility contests. Yet for Sartre and other existentialists the very idea that the self needs a Thou in order to be is an expression of bad faith. Hell is other people, as Sartre put it in his play *No Exit*. Our relations to others must be specified with respect to the demand on individuals to confront the fact of their existence, to bear the weight of freedom.

Thus, the idea of responsibility is embedded in the modern debate about how to understand human existence. It can be given different definitions in ethics and these differences come to expression in how morality and moral agents are understood. Theories of responsibility in fact focus on different senses of the Latin root *respondeo*. Agential theories of responsibility, as I have called them, elaborate the meaning of this concept in terms of accountability. Social theories examine the importance of representative action in the moral life. And those theories which I have called dialogical center ethical reflection on the root phenomenon of answering. The discourse of responsibility allows us to chart the similarities and differences between modern forms of ethics with respect to the question of moral identity. The important connection between the language of responsibility and theoretical options will be developed more fully when we examine theories of responsibility.

However, the discourse of responsibility is more complex than I have presented thus far in this chapter. The modern idea of responsibility also owes a historical debt to classical ethics and to religious traditions, especially the Christian tradition. If we are to understand the idea of responsibility and sharpen our conceptual tools, we must trace these connections as well.

THE TRADITIONAL IDEA OF RESPONSIBILITY

As we have seen, responsibility is a late-comer in Western ethics. Yet, Aristotle, in the *Nicomachean Ethics*, provides a discussion of something like the modern idea of responsibility.[13] His analysis shaped subsequent ethics. If we are to grasp the full range of conceptual issues surrounding the idea of responsibility, it is important to explore Aristotle's ethics. This is especially the case with moral freedom and also praise and blame.

In Book III of the *Nicomachean Ethics*, Aristotle focuses his discussion on the voluntary (*hekousion*) and involuntary (*akousion*) as necessary for determining when it is right to praise or blame a person for his or her actions. It is this problem that leads Aristotle into the question of the nature of voluntary actions. He begins with a social practice of praise and blame and then examines the domain of freedom. Further, in contrast to most modern ethics, his conception of the voluntary does not mean unfettered choice or choice based on rationally universalizable principles. The moral person, he insists, acts under the rule of the so-called "golden mean." Given the kind of person an agent is (say, shy or prideful) and the circumstances of action, the agent in making a choice must seek a mean between excess and deficiency (the shy person, for example, needing more courage than the prideful one in the same situation). The "norm" of choice is thus formally the same in all situations (i.e., the mean), but it is materially different with respect to the agent and circumstance. Moreover, despite differences of temperament and moral rectitude, Aristotle reasons that we are moved to choose or avoid choosing due to what is found to be beautiful or ugly, advantageous or hurtful, pleasant or painful. In each case we are to seek the "mean" between excess and deficiency in our choices.

For Aristotle, our acts of choosing are always matters of desire, want, inclination, and evaluation. We do not choose blindly or based solely on abstract principle; we choose with respect to what moves us to act. This is why the principle of right choice (the mean) can be articulated formally, but must also be understood substantively with respect to the actual situation of particular agents choosing, agents characterized by different capacities, dispositions, and character traits. As Aristotle puts it:

Virtue, then, is a state of character concerned with choice, lying in a mean, i.e., the mean relative to us, this being by a rational principle, and by that principle by which the mean of practical wisdom would determine it ... Hence in respect of its substance and the definition which states its essence virtue is a mean, with regard to what is best and right an extreme.[14]

Aristotle's concern is to determine the proper grounds for praise and blame and thus to determine merit and demerit. This is based on a conception of virtue and also, as we see below, what defeats assignments of responsibility. The idea of "freedom" is thus set within these other moral concerns.

This approach to the question of moral freedom is important for two reasons. First, Aristotle's account of the "voluntary," as Marion Smiley notes, means that the ethicist does not need to establish the existence of a "self" who chooses abstracted from a community and its practices of praise and blame. This pattern of moral reflection stands in contrast to the enterprise of much modern ethics which, as we saw above, begins with the freedom of individual agents as the condition for understanding the constitution of the social contract. Second, Aristotle "does not conceive of blameworthiness as a black spot on one's soul. Instead, he conceives of it as part of our social practice of blaming."[15] We can make sense of praise and blame, or, more generally, the assignment of responsibility, by exploring social practices without thereby needing to isolate a place *in* the agent (call it the soul) to which praise and blame are somehow attached. As we will see shortly, this stands in contrast to Christian conceptions of responsibility and personal inwardness.

What then does Aristotle mean by voluntary action? An action is voluntary when the agent is the spring or cause of the action, even if that "cause" is the desire for the pleasurable; it is involuntary, he contends, if the cause is external to the agent. As he writes,

Since that which is done under compulsion or by reason of ignorance is involuntary, the voluntary would seem to be that of which the moving principle is in the agent himself, he being aware of the particular circumstance of the action.[16]

Aristotle analyzes voluntary action in terms of freedom from constraint or compulsion and ignorance. Voluntariness is the condition for responsible action; it entails, as Aristotle indicates, the ability to act knowingly in specific circumstances. The focus of his analysis of the voluntary is on the assignment of responsibility, and, correspondingly, on the ways in which assignments of responsibility can be defeated. As J. R. Lucas has noted, "Aristotle reckons that the ascription of responsibility for an action can be defeated in two ways: 1. that what was done was done by compulsion ... (*bia*), or 2. it was done through ignorance ... (*di'agnoian*)."[17] And Aristotle draws a further distinction between action done through ignorance (*di'agnoian*) and action where the agent is being ignorant (*agnoon*). An agent cannot appeal to ignorance of an action's being wrong since, if allowed, that appeal would defeat every assignment of responsibility. Rather, the agent can defeat responsibility assignments by showing that she or he was ignorant, maybe necessarily so, of conditions or circumstances which affect the morality of an action.

Aristotle's analysis of what we would call responsibility centers then on questions of agent causality and the conditions for rightly assigning responsibility. The norm of the moral life is happiness (*eudaimonia*) which is the exercise of the faculties in accordance with virtue or excellence.[18] What is more, Aristotle does not interpret the formation of moral character through the idea of responsibility. He conceives of moral character in terms of the virtues, moral and intellectual. The Aristotelian theory of responsibility is a "weak" theory which centers on the agent/act relation in determining the moral rightness of social acts of assigning responsibility, that is, praising and blaming. It focuses attention on questions of freedom and responsibility as well as on what defeats assignments of responsibility.

This same pattern of thought was followed in Christian theology by Thomas Aquinas. To be sure, Aquinas reconstructs Aristotelian ethics in terms of the doctrine of virtue, to which he adds the theological virtues, and in terms of natural law and the conception of the highest good (*summum bonum*), the God of Christian faith. Aristotle would find the theological virtues of faith, hope, and love unintelligible. These virtues are infused into the human soul through the Holy Spirit and thus are not gained by human effort (i.e., learning and habituation) as is the case with the moral and intellectual virtues. God is the sole cause of the theological virtues even as they enable the person to act in meritorious ways.[19] In this respect, an external cause becomes an internal principle of action *in* the agent. The theological virtues are necessary for persons to attain the highest good of the vision of God.

Granting these theological revisions of Aristotelian ethics, Aquinas begins his treatment of human acts with the problem of the voluntary and the involuntary. In the *Summa Theologiae* he writes:

There must needs be something voluntary in human acts. In order to make this clear, we must take note that the principle of some acts is within the agent, or in that which is moved; whereas the principle of some movements or acts is outside ... For since every agent or thing moved acts or is moved for an end ... those are perfectly moved by an intrinsic principle whose intrinsic principle is one not only of movement but of movement for an end.[20]

Once again, moral inquiry focuses on what moves an agent to act and the location of the principle of action. If the principle of action is internal to the agent, no matter what that principle may be (say, reason or passion), then the action is voluntary. The main idea is that an agent has control over his or her conduct in so far as he or she is the principle of action. This has been a claim basic to Christian ethics. St. Augustine, for instance, argued that

Our will would not be a will if it were not in our power. Because it is in our power, it is free. We have nothing in our power which is not free. Hence it is not necessary to deny that God has foreknowledge of all things, while at the same time our wills are our own.[21]

Christian theologians have insisted that, while God foreknows all future actions and states of affairs, persons are none the less morally free. The answer to this glaring paradox has to do with the extent to which the principle of choice is *in* the agent, even if, it is claimed, God foreknows what the choice will be. If that principle is *in* the agent, then the agent is free. Because of the fall, I always and necessarily act in sin, but in so far as the principle of action, my orienting love as Augustine puts it, is *in* me, then I choose freely.

Thus, in traditional Christian ethics certain conditions must be met in order for agents to have some power over their lives and actions. Voluntariness is defined in contrast to compulsion rather than as freedom from internal inclination, desire, or interest. One acts voluntarily even if internal passion for some good overrides better judgment. This is the case for instance with vicious actions. There are also epistemic conditions for voluntary action: ignorance of circumstances (but not of principles) can render an agent's action involuntary, say due to deception. It is not surprising, then, that Aquinas concurs with Aristotle about what defeats assignments of responsibility. Compulsion and ignorance render an action "involuntary" and thus morally not subject to praise and blame.

Aquinas and Aristotle represent the traditional Western conception of responsibility. This account centers on knowledge and the capacity or power to act as necessary conditions for voluntary action, and, also, on the reasons why responsibility might not be rightly assigned or imputed to an agent. These themes continue to dominate ideas of responsibility in Western cultures. In Chapters 6–7 below I will return to these matters, and, in fact, develop a theory of freedom indebted to these classical thinkers. At this juncture we must further clarify the conceptual field of the idea of responsibility by turning to yet another historical source, the Christian tradition.

RESPONSIBILITY AND CHRISTIAN INWARDNESS

The connection in modern ethics between responsibility and a host of other moral concepts, like conscience and obligation, owes a debt to the Christian tradition. Judith Shklar, for instance, contends that modern conceptions of moral responsibility are "perfectly meaningless unless one believes in sin." She continues:

To understand at all the emphasis on the interiority of morals one must remember traditional Christian morality, from which it is derived and which alone makes it intelligible. To speak of this internality as if it were an obvious psychological or social fact is absurd and has strange results in practice. For orthodox Christians, now as ever, the most extensive category of human failure is sin. A sin is not just injury to another person. It is rejection of God. It may be a purely internal act which may lead to no change in external behavior.[22]

The idea that individuals are morally responsible outside of particular social practices of praising and blaming requires a conception of someone to whom they are responsible, someone or something that transcends given social practices. It seems to require a conception of God. And it also requires a specific idea of human inwardness which the ideas of conscience and responsibility seem to articulate. The problem with a good deal of modern ethics, Shklar laments, is that it denies or forgets its origins in Christian thought and therefore fails to escape that origin. Despite the rhetoric of secularity in most modern ethics, the very concept of obligation and the idea of the self betray Christian origins.

Shklar and other critics of modern conceptions of the inwardness of life are correct to note that the interiority of the moral life is reflected in Christian discourse about conscience and the moral standing of persons before God. The Christian tradition has consistently insisted on the interiority of human life, an inwardness not subject to cultural, social or ideological tyranny, but related to the divine. The *locus classicus* for this outlook is found in Romans 2:13–16 which in turn refers to Deuteronomy 10:16 and Jeremiah 31:33–34. In Romans St. Paul writes:

For it is not the hearers of the law who are righteous in God's sight, but the doers of the law who will be justified. When the Gentiles, who do not possess the law, do instinctively what the law requires, these, though not having the law, are a law to themselves. They show that what the law requires is written on their hearts, to which their own conscience also bears witness; and their conflicting thoughts accuse or perhaps excuse them on the day when, according to my gospel, God, through Jesus Christ, will judge the secret thoughts of all.

The tribunal before which the self is summoned to appear is conscience, upon which is engraved the law of God. Before this tribunal the self appears as convicted, as under the law, or as having its righteousness in Christ. Norms of responsibility are not reducible to individual choice, preference, or even social practices and conventions. The norms of responsibility are grounded in the will of God discovered through the examination of conscience.

Christian faith has contributed to the modern sense of inwardness and also to the belief in the dignity of persons. These are convictions we must never abandon. I develop an account of moral identity and conscience in Chapter 7 to make this point. However, Shklar is incorrect to claim that the idea of responsibility depends solely on Christian thought. Marion Smiley has argued that despite their differences, both classical Greek and traditional Christian thought "posit a relationship between individuals and an external blamer – the community and God, respectively."[23] In Aristotelian ethics the community and its social practices of praise and blame are the focus of questions of moral responsibility. Within traditional Christian thought God is the ideal blamer, as Smiley puts it. What is at stake in classical moral outlooks is the way in which persons are held responsible for their actions. The grounds for assigning responsibility are objective to the self. Modern ethics has moved the problem inward as part of the processes of interiorization and individualization that we have been tracing. Either the self becomes its own ideal blamer, is responsible for itself in the quest for authenticity through conscience (Lévy-Bruhl) or an existential act of choice (Sartre), or the social values which help to define a person's way of life are the internalization of communal mores (Bradley). The agent is ultimately responsible for herself or himself, whether or not this is understood in existentialist terms. And this makes the demand of authenticity, to be true to self, basic to the moral life.

The idea of responsibility provides an entrance point for examining strands in the Western moral heritage. Yet it also suggests that an adequate account of responsibility cannot deny the historical influence of theological discourse on moral consciousness. Modern ideas of responsibility draw on and react against the theological background of Western moral discourse. The importance of this point should not be missed. The loss of a theological component in understanding the human condition is, of course, definitive of the modern world. If Shklar is right this has freed us from destructive forms of guilt, which is, as I noted in Chapter 1, a deep aspiration in contemporary thought. At least it has meant that guilt is now a matter of social practices and not our standing before the divine. But Shklar's and Smiley's criticisms of Christian thought are not the whole story. Questions of sin, guilt, redemption, and forgiveness actually rest on deeper claims about human existence. What do I mean?

In general, modern Western thought denies that human beings are created in the image of God, whether or not that "image" is defined through the connection between conscience and moral responsibility. This denial of the intrinsic dignity of the self signals a shift in how to understand moral responsibility. Hans Jonas has astutely noted that

> the paradox of the modern condition is that this reduction of man's stature, the utter humbling of his metaphysical pride, goes hand in hand with his promotion to quasi-God-like privilege and power. The emphasis is on *power*.[24]

Modern ethics does not understand the human as the *image of God*, but as a product of evolutionary processes, social relations, or discrete acts of freedom. This means that the central moral question becomes the status of power. Even the person is understood in modern thought as the origin of action, and thus a source of power. And this raises basic moral questions. Does power determine what is right such that the strong ought to dominate the weak? In political associations are norms contractually grounded in order to stop the war of each against each found in the state of nature, as Hobbes and others have argued? Is it possible, as some philosophers contend, to speak of human rights without appeal to human nature and dignity, but simply in terms of capacities for action?[25]

Late-modern societies continue to believe that power does not determine the standard of right and wrong, and also to insist on some idea of human inwardness and dignity. This fact indicates the way in which contemporary ethics remains indebted to inherited religious and also premodern philosophical convictions. It has enabled some theorists, like Smiley, to address problems in the modern, liberal conception of responsibility by retrieving classical ethical positions, for instance Aristotelian ethics. What has not been attempted is a reconsideration of the theological dimensions of the question of responsibility in light of the question of power.

From this brief analysis of the semantic and conceptual field of responsibility it is clear that this idea covers distinct but related matters in ethics. And we have also seen how the history of the discourse of responsibility brings to light developments in Western culture. Based on this analysis, we can now specify features of responsibility. These are the linguistic and conceptual building blocks for an integrated theory of responsibility.

THE FEATURES OF RESPONSIBILITY

Within the discourse of responsibility several features of human action are made evident. A person, or community, *bears* responsibility for something such that the action can be ascribed to the agent(s). One can also *assume* or take responsibility for oneself or for another as well as *ascribe* it to others. Likewise, we speak of responsibility in terms of being called to *account* for something. Persons must also *answer* for their actions to others and to themselves. These diverse meanings center on the assignment of responsibility (bearing, assuming, ascribing) as well as the basic phenomena of accountability, answerability, and representative action. It is also the case that these senses of responsibility are factually independent. As Roman Ingarden writes:

One can be responsible but neither called to account, nor assume responsibility ("take it upon oneself," as Nicolai Hartmann puts it). And conversely, one can be called to account for something without being in fact responsible for it. One can also assume responsibility for something, without being actually responsible for it. Given that someone is responsible for something, he should both assume responsibility and be called to account for it.[26]

The conceptual problem for ethics is to distinguish and yet relate these diverse features of responsibility. That is, an adequate theory of responsibility must interrelate the phenomena of accountability, answerability, and representative action and distinguish the practices of ascribing responsibility to other agents and assuming responsibility for oneself. This is because the idea of responsibility designates in formal terms what it means for human beings to relate to others and the world as agents. To be a moral agent is to be responsible *for* oneself, and perhaps others, in and through responding *to* others and being accountable *for* bringing something into being or acting on behalf of others through the exercise of power.[27] Each feature of responsibility is important and needs to be clarified.

Accountability is basic to assigning culpability to persons, to establishing the grounds and the limits of praise and blame. The language of moral accounting, as William Kneale notes, "began with an extended use of debt words to cover the whole range of duty."[28] The idea that persons are liable for their debts to others and must account for them is the connection between account-ability and responsibility. The focus here is on the agent/act relation. And this implies conditions for ascribing responsibility. We are accountable for our character and conduct and the larger social and cultural world in so far as we help to bring them into being. Much of what is ordinarily meant by responsibility is simply accountability for actions. Debt language also permeates religious discourse. For instance, St. Anselm, in a classic treatise on the atonement, argued that sinful human beings owe a debt to God, a debt they cannot pay but which God does pay through Christ. Moral accounting and the practices of praise and blame take on ultimate significance in this theological framework.[29]

Answerability denotes the responsive dimension of responsi-bility. We are answerable *to* someone or something, including ourselves, that has a rightful claim or authority to determine standards of conduct. That which summons an answer might be another person, a community, the State, or, in classical thought, the "voice" of conscience and the dictates of God engraved on the human heart. We are creatures who must answer for our lives with respect to what or who questions us.[30] The focus of moral attention is then not the agent/act relation, but the self/other encounter. The dimension of answerability concerns the relations, norms, and values by which we ought to make decisions as these are "heard" by the self. It is little wonder, then, that theologians have traditionally asserted the need to "hear" divine commands and have spoken of the "voice" of conscience. Above we found that this idea was already present in St. Paul's letter to the Romans.

Finally, we are accountable *for* what we do and are answerable *to* something or someone because we are responsible *for* ourselves and in certain situations for others as well through representative action. This dimension of responsibility denotes the moral identity of an individual or a community. Identity is imputed to an agent or community in a variety of ways, often through narratives we tell or adopt, declarative statements ("I ate the banana"; "She insulted Harry"), or by stipulation ("The owner of an ox which gores another shall pay reparations").[31] "Thou are the man," Nathan said to David about the murder of Uriah (2 Sam. 12:7). Designating "who" acted through the assignment of responsibility in whatever linguistic form it takes identifies an agent accountable and answerable for himself or herself. Moral identity, responsibility for self, draws together the other features of responsibility. Yet moral identity is deeply social since it is linked to responsibility assignments and also to the power or freedom to act in the world. This is why, we might imagine, matters of assigning responsibility, the validity of praise and blame, always pose other questions about voluntary action.

This connection between freedom and responsibility in terms of moral identity, a theme we will explore further in Chapters 6–7, requires a complex idea of freedom correlate to features of responsibility. In so far as we designate "who" is acting, moral freedom entails, but is not limited to, the capacity to make choices for which we are accountable. To say "who" is accountable, and, thus, open to praise and blame, requires that some person or community actually act, exert force in the world subject to moral norms and values. But moral identity is not simply a description of the act or intention to act. Freedom is also the condition for answerability. We only ask "who" is acting and also answer for our actions or for the actions of others if basic conditions necessary for moral action are evident, that a person can consider what to do and has the ability to do it. This is what gives a sense of depth to human beings; we have capacities for action and response to others, and thus are culpable for our actions, and these capacities undergird our actual choices. But the fact that someone can, in principle, live by a moral ought and answer for actions and intentions does not yet tell us "who" she or he is. Moral freedom is most basically, I argue, the power to interpret "who" one is with respect to actual life. The freedom to be responsible *for* ourselves, and also for others, entails the other features of responsibility and the forms of freedom they manifest. This relation between the forms of freedom and moral responsibility will concern us throughout the rest of the book.

CONCLUSION

This chapter has explored the semantic and conceptual field of the idea of responsibility. Responsibility is a complex moral idea that provides the resources for exploring and interrelating dimensions of the moral life. Thus, I have developed in this chapter the tools for reconstructing the ethics of responsibility from a theological perspective. The task of the next chapter is to move from the analysis of language and concepts to the examination of theories of responsibility. It will then be possible in Chapter 5 to outline a theory of value and the imperative of responsibility of an integrated ethics of responsibility.

CHAPTER 4

Theories of responsibility

This chapter has two purposes. First, I provide a typology which has broad application to theories of responsibility. Second, the chapter examines moral theorists important for the constructive argument of the book. In Chapter 2 above, I outlined two ways in which to develop a typology of theories of responsibility. The typology of this chapter is developed around those distinctions. I examine agential, social, and dialogical theories and then explore the difference between strong and weak positions within each type.

AGENTIAL THEORIES OF RESPONSIBILITY

Agential theories of responsibility focus on the acting agent. They determine the rightness of acts of praise and blame with respect to the connection between the agent and her or his deed. If it can be shown that an agent did not "cause" an act to happen, or the the agent while causing the act had excusing reasons, then the agent is not morally responsible. Agential theories differ with respect to the principle for judging acts moral or immoral. If a principle of moral choice and judgment is claimed to be valid only when grounded in the self-legislating capacity of the agent, then the theory is a strong, agential theory of responsibility. This follows since responsibility for self, that is, moral autonomy, is the first principle of ethics. Conversely, if valid moral principles are grounded in something other than the agent, say in social practices or the will of God, but the theory still holds the agent/act relation to be basic to the idea of responsibility, then the theory is a weak, agential account of responsibility. We can clarify these distinctions by examining examples of agential theories of responsibility.

Autonomy and moral responsibility

Undoubtedly the most important modern expression of a strong, agential theory of moral responsibility is the work of Immanuel Kant. To be sure, Kant does not use the term responsibility in his ethics. Yet in so far as autonomy, and thus responsibility for self, is basic to his ethics, it is fitting for us to consider Kant's moral philosophy as an ethics of responsibility. Moreover, my integrated theory of responsibility draws insight from Kant's insistence on the centrality of respect in the moral life.

Kant's ethics centers on determining the supreme principle of morality. He calls this the categorical imperative. The categorical imperative is the principle of non-contradiction applied to the will. As Kant notes in one formulation of the imperative, "Act always on such a maxim as thou canst at the same time will to be a universal law; this is the sole condition under which a will can never contradict itself; and such an imperative is categorical."[1] The imperative is categorical in that it represents an action as necessary of itself without reference to ends or consequences. It is not surprising, then, that Kant focuses his attention on the will. Because an action is right or wrong in virtue of its objective necessity and not its consequences, the moral agent must act purely from obedience to the imperative. Imperatives, Kant notes, "are only formulae to express the relation of objective laws of all volition to the subject of imperfection of the will of this or that rational being, for example, the human will."[2] In Kant's ethics, the supreme principle of morality is thus correlated to an account of the moral agent. Ethics is not grounded in the empirical claims of anthropology. Nevertheless, the ethics provides an account of what it means to be a moral creature.

The burden of Kant's ethics is to show that, while human beings are moved by desire and inclination and subject to the causal laws of the universe, this is not all that can or must be said about the agent. If that were so, then determinism would be true. The central moral problem of how to govern our lives would not be a real problem. Kant wants to show that a person can legislate maxims for her or his actions. In so far as this is the case, then we must postulate freedom as the necessary condition for morality. Freedom is not experienced in the world; it is not a phenomenal reality like the physical world around us. It is an idea for the noumenal reality of human beings; it is a postulate of pure practical reason to specify the conditions of the possibility of our being moral agents.

The supreme principle of morality and the postulate of freedom are interlocking ideas. Not only must an agent will that a maxim of her or his action be a universal law of nature, but she or he must also respect the humanity in persons as members in a Kingdom of Ends. This is because, Kant writes, "Rational nature exists as an end in itself."[3] A person, Kant reasons, apprehends her or his existence as an end in itself and not simply as a means to something else. Under the demand to universalize our maxims of action, we must then acknowledge this about others: persons are never to be used only as means to other ends. And, since the idea of humanity specifies the object of respect, it means that all persons are members in the Kingdom of Ends.[4] Now in formulating the categorical imperative one must postulate freedom. That I *ought* so to act, Kant holds, means that I *can* so act. This requires that "the basis of obligation must not be sought in the nature of man, or in circumstances in the world in which he is placed, but *a priori* simply in the conceptions of pure reason."[5] What obligates or necessitates the will is an a priori concept of reason, since an account of the world or the "nature" of man, as Kant puts it, shows us not to be free. But how can a concept be the basis of obligation? Kant's answer is developed through the deduction of freedom as the basis of obligation, the formulation of the law of freedom as the categorical imperative, and, finally, a specification of the moral good. We must examine this in more detail.

If moral agents are to be genuinely free, they must legislate their own actions. In order to deduce freedom, Kant begins the *Fundamental Principles of the Metaphysics of Morals* with the idea of the good will, a will which acts from duty alone, and thus the idea of duty. "Duty is," he writes, "the necessity of action from respect for law."[6] This suggests that what ought objectively to determine the will is *law* while its subjective determination is *pure respect* for this practical law. Kant's argument hinges, then, on finding the unity of this law and respect in the will, in practical reason, as the law of freedom itself. In order to understand this we must grasp the peculiar status of the *feeling* of respect.

If the feeling of respect were like all other natural feelings, then Kant's ethics, despite his protests, would be grounded in the empirical nature of the human, and, thus, would not find its basis in rational freedom. But Kant argues that the feeling of respect is "*self-wrought* by a rational concept."

The immediate determination of the will by the law, and the conscious-ness that my will is *subordinated* to this, is called *respect*, so that this is regarded as an *effect* of the law on the subject, and not a *cause* of it. Respect is properly the conception of a worth which thwarts my self-love ... The *object* of respect is the *law* only, that is, the law which we impose on *ourselves*, and yet recognize as necessary in itself.[7]

The feeling of respect discloses the legislative power of practical reason through a conception of a worth that thwarts self-love. This is simply another way of saying that freedom and the moral law are coordinate in Kant's ethics. Not surprisingly, Kant holds that the only good without qualification is the good will.

What does this have to do with responsibility? As we have seen, theories of responsibility draw a connection between causal judgments and evaluative judgments. Agents are praised or blamed with respect to what they have directly or indirectly caused to happen. Kant's ethics is no exception to this rule. Indeed, he seeks to show the way in which a moral agent is a free cause in the world. But the question of responsibility, given Kant's construal of freedom, is whether or not an agent is responsible for the maxim on which she or he acts. An agent is only secondarily responsible for the consequences of an action; she or he is properly responsible, morally praised and blamed, with respect to the motive for acting.

Kant's ethics is a strong, agential theory of responsibility because the self-legislating freedom of the agent and the principle of that legislation, the categorical imperative, is the first principle of ethics. But this means, oddly enough, that an agent is not responsible for promoting any specific good in the world. She or he is only responsible for promoting the goodness of the will. The purpose of morality, Kant insists, is to make *oneself* worthy of happiness. In terms of an integrated theory of responsibility, Kant has articulated the demand of respect for persons. Yet Kant has not rightly addressed the demand to enhance or promote the integrity of all of life, but just the goodness of the will.

Theonomy and moral responsibility

Theologians earlier in this century were not troubled by the Kantian constriction of moral concern to the rectitude of the will. What theologians contested was Kant's contention that theological ethics was necessarily *heteronomous* because the person is bound by a foreign will, the will of God. In order to counter this charge, Paul Tillich used the idea of *theonomy* in developing a Christian version of an agential theory of responsibility. Tillich argued that the moral law, the law of our essential nature, is not simply something we impose on ourselves (autonomy), nor is it simply imposed on us by another, foreign will or power (heteronomy). The Christian message is that true freedom is theonomous; the moral law of God is nothing else than our true being. What does this mean?

Tillich makes several arguments of importance for our typology of theories of responsibility. First, he argues that the "moral imperative is the command to become what one potentially is, a *person* within a community of persons."[8] The moral act aims at actualization of self, that is, constituting the person as person. "For the ethical problem this means," Tillich writes, "that the moral act is always a victory over disintegrating forces and that its aim is the actualization of man as a centered and therefore free person."[9] The experience of the imperative to actualize the self is found in conscience, the silent voice, as Tillich puts it, of our essential nature judging our actual lives. The moral law is nothing else than our essential nature formulated in terms of an imperative. It is not heteronomous. To act on this law is to act out of freedom since it is to act according to our essential being.

The next point in Tillich's ethics is to show that this law, this formulation of essential being as obligatory, is simply the will of God. He writes:

The "Will of God" for us is precisely our essential being with all of its potentialities, our created nature declared as "very good" by God, as, in terms of the Creation myth, He "saw everything that he made." For us the "Will of God" is manifest in our essential being; and only because of this can we accept the moral imperative as valid. It is not a strange law that demands our obedience, but the "silent voice" of our own nature as man, and as man with an individual character.[10]

In distinction from Kant, Tillich argues that the validity of the moral law depends on its grounding in human nature. But the "nature" in question is not, as Kant thought, our empirical desires, wants, and inclinations. It is the essential being of the human. The moral law articulates what we most essentially are, centered persons in relation to God as the ground and power of being, as an imperative for how to live amid the ambiguities of actual life. In so far as our existence is a matter of concern to us, we confront our being under an unconditioned demand. The symbol for the ground and power of the unconditioned is "God." The "will" of God is nothing else than the good of essential being. True freedom is then theonomous; it is not something we legislate but is grounded in God. Put differently, for Tillich, "Morality does not depend on any concrete religion; it is religious in its very essence."[11]

However, the fact is that our actual being is not our true being. Our lives are marked by fragmentation. This is Tillich's other basic point in ethics. And it is why we encounter the moral law, the will of God, as an imperative for action; it is the testimony of conscience against the self. "The voice of man's essential being is silenced, step by step; and his disintegrating self, his depersonalization, shows the nature of the antimoral act and, by contrast, the nature of the moral act."[12] How is this ethical problem – the conflict between self-integration and disintegration – to be answered? For Tillich, the answer is found in love, *agape*, as the ultimate principle of morality. *Agape* "points to the transcendent source of the content of the moral imperative" and unifies our actual nature with our essential being.[13] Love answers the moral problem.

As Dietz Lange has noted, Tillich conceives of love as the reunification of estrangement, the actual with the essential. Lange notes that

The fundamental problem of ethics is the unity of being; ethics is then the question of the self-integration of the human in its existence determined by the polarities of freedom and destiny, dynamics and form, individualization and participation. Self-integration presupposes the centering of the human around its essence, its determination.[14]

Agape overcomes the estrangement between fallen existence and the goodness of created, essential being. This love, *agape*, also draws within itself justice, as the acknowledgement of the other person as a person, and the power to act. Tillich's ethics stresses, then, the demand for the actualization of life against disintegrating forces. He conceives of this in terms of the reunification in *agape* of actual life with essential being. By acting on the moral law in obedience to our essential nature a higher mode of being is actualized, that is, the person in community with other persons.

Tillich presents a Christian agential theory of responsibility. An agent is responsible for self-actualization under the law of essential human nature. Yet Tillich does not fundamentally alter the Kantian conception of responsibility for self. What he does change, significantly, is the conception of the moral good. The moral good is not simply the good will acting out of respect for the moral law; it is the actualization of the person as a multidimensional reality. In this respect Tillich voices a concern of an integrated ethics of responsibility. As we have seen, this ethics insists on the proper integration of the multidimensionality of life and its various goods. Yet because Tillich conceives of this integration in terms of the relation between the essential and actual self, his real concern is not the integrity of diverse goods in historical and social life. This means that his ethics verges on intuitionism in the appeal to the "silent voice" of conscience about what to do, rather than examining the range of questions which constitute the field of morality in terms of the actual values and disvalues of life.[15] At this level, Tillich's *agapistic* ethics differs from an integrated ethics despite a similar concern for the integrity of life.

Agential theories of responsibility

What have we seen by examining agential theories of moral responsibility? These theories focus moral analysis on the relation of the agent to himself or herself. They attempt to specify the grounds for validly relating causal and evaluative judgments with respect to the capacity of the agent to act morally. A valid moral law is an expression of the self-legislating power of practical reason (Kant) or the law of our essential nature (Tillich). This means, first, that an agent is not *morally* responsible for duties and obligations specified on other grounds. An agent might have political, economic, or social responsibility for certain acts, but the moral character of responsibility is not so grounded. In assignments of moral responsibility, an agent validly assumes and others rightly ascribe responsibility to him or her only when the action expresses or violates the ground of the moral law (rational freedom; essential nature). Second, the assignment of responsibility is a secondary activity dependent on the defining act of moral existence itself – autonomous or theonomous existence. In this respect, agential theories of responsibility reverse the line of reasoning found in the traditional, Aristotelian theory of responsibility. In these theories, freedom is the necessary condition for making sense of the assignment of responsibility.

By concentrating on the validity of moral claims with respect to moral freedom, strong, agential theories curtail moral consideration of consequences in assigning responsibility (Kant) or the actual goods which can and ought to be integrated in life (Tillich). From the perspective of an integrated ethics of responsibility we have reason to question the constriction of moral evaluation to the agent/act relation in strong, agential theories. In fact, this point is what distinguishes strong from weak agential theories. In assigning responsibility, a weak theory might argue that agents ought to pursue goods (Aristotle's virtue ethics) and that they are responsible for the consequences of action with respect to other goods (in utilitarian ethics). These arguments would require one to show why an agent is so bound. In order to show that, one must specify some claim about what grounds the agent/act relation. Thus, a strong theory seems required even to make sense of other goods (for example, virtue) or a utilitarian assessment of consequences. As we will see in a moment, social theories of responsibility answer this question on the grounds of social practices and thus negate the basic claim of agential theories.

Kant and Tillich articulate insights necessary for an integrated ethics of responsibility. Kant rightly insists on respect as basic to morality. The feeling of respect is the recognition of a worth which thwarts self-love. Similarly, Tillich rightly stresses the demand for the actualization of the integrity of life against disintegrating forces on life. While he conceives of this in terms of the reunification in *agape* of actual life with essential being, the insight that morality is about the integrity of life is obviously basic to an integrated ethics of responsibility. In addition, these thinkers argue that by acting on the moral law or in obedience to our essential nature in love a higher mode of being is actualized, i.e., moral autonomy (Kant) or the actualization of life (Tillich). Strong, agential theories of responsibility provide backing for the claim that by acting under the imperative of responsibility a higher form of life, a unique good, is enacted in life, the ethical good of moral integrity.

Such are the insights and oversights of the first type of theory of responsibility found in ethics. An integrated ethics incorporates the insights of this type of theory while also addressing our intuitions about morality not found in this form of ethics. In order to complete our analysis we must examine the other types of theories of responsibility.

SOCIAL THEORIES OF RESPONSIBILITY

Social theories of responsibility focus on social roles, vocations, stations, and thus communal unity. As I noted in Chapter 3, F. H. Bradley in his *Ethical Studies* was an early exponent of this type of ethic. For Bradley, and other social theorists, "an individual human being, in so far as he or she is 'the object of his [or her] self-consciousness,' is characterized and penetrated 'by the existence of others.' In short, the content of the self is a pattern of relations within a community."[16] In order to realize one's existence in the world, one must assume one's proper place in the social world. Moral duty flows from a person's social function which comes from social roles and station.

The basic claim of social theories of responsibility, in Peter French's words, is that moral persons "come into existence at various levels of description or, more to the point, via descriptions."[17] Personal identity does not exist prior to descriptive practices, but comes to be through those practices, practices like praise and blame and hence the assignment of responsibility. If we can specify the rules for description, or, more properly, the rules for praising and blaming agents, we can fulfill all that can or need be said about conditions for moral agency and responsibility. I can clarify this type of theory now by exploring important representatives of it.

Responsibility and community

The philosopher Marion Smiley has recently proposed a pragmatic, social theory of responsibility. She traces the deep roots of this account of responsibility not to Bradley but to Aristotle. As we saw in Chapter 3, for Aristotle responsibility is (in part) a matter of the social practice of praise and blame. Smiley argues that even Aristotle's discussion of the voluntary is "concerned primarily with the conditions under which we as social and political blamers consider an individual's actions to have been voluntary or involuntary."[18] This is a boon for Smiley since her intention is to move discourse about responsibility beyond modern and Christian conceptions which rely heavily on a notion of the self and internalized transcendence. The person held accountable before, and blamed by, God becomes in the modern world its own ideal blamer. The Aristotelian position, on Smiley's reading, is a way beyond the modern conception of the responsible self.

The focus of Smiley's theory is on moral responsibility rather than the problem of free will and determinism found in much ethical theory, especially, as we have seen, in Kantian ethics.

While Smiley grants that in fact assignments of responsibility, that is, evaluative judgments about conduct, are bound to causal judgments about who brought about something, she contests the claim that evaluative judgments are determined by, or reducible to, causal ones. We often rightly hold persons responsible for actions they did not directly cause. This is because practices of praise and blame are keyed to social roles. The head of a surgical team by virtue of her social role can be held responsible for an action she did not directly cause. To be sure, something had to have happened, and in this sense there is a causal judgment to be made in assigning responsibility, but a causal judgment does not in itself determine the rightness of assigning responsibility.

Smiley views "moral responsibility as part of social and political practice rather than as an ideal that supposedly exists outside and superior to our social and political concerns."[19] The meaning of responsibility is tied to what specifies moral blameworthiness and not only to the problem of free will. This departs from those forms of ethics, seen in strong, agential theories of responsibility, which isolate something internal to self – practical reason, soul, or conscience – before and to which the agent is responsible. Responsibility is "a judgment that we ourselves make about individuals on the basis of our own social and political points of view."[20] The source of responsibility is the practice of blaming and not the individual's free will that caused an event. We might put it like this: Smiley contests the assumption that responsibility is something we *discover* about individuals rather than *assign* to them. She challenges the idea that we can or ought "to conflate our practical judgments of causation and blameworthiness into one ostensible factual discovery of moral responsibility."[21] Smiley contends, conversely, that responsibility is precisely a practice of assigning praise and blame keyed to causal judgments, social norms, and the roles of agents.

The practice of blaming is pragmatically understood by Smiley in terms of how a community or individual responds to problems and crises within the social world. The social practice of blaming provides means to control the social world and also to form the character of agents. As Smiley notes, "blaming as a practice is necessary to the construction by individuals of a relationship between them and the external world, a relationship which partly defines, and in turn helps to maintain, their personal integrity."[22] Following the work of Bernard Williams, she defines personal integrity in terms of projects to which an individual is committed as what life is about.[23] Responsibility and the social practices it entails are commitments basic to the integrity of the self. But the self, according to Smiley, arises within these social practices. We do not discover the self as the necessary condition of free action; responsibility is not dependent on the inner self holding the acting self accountable. Rather, we speak of the self with respect to social practices, including the practice of blaming. In doing so, we are defining and maintaining individual and collective integrity.

On reaching this conclusion, we are at the heart of Smiley's moral theory. By taking practical judgments about responsibility seriously, she hopes to use them "to open up channels of communication with regard to the moral responsibility of particular individuals for the suffering of others."[24] This also entails a shift of moral attention away from responsibility for action and character to the problem of external harm. Smiley grants that moral situations are precarious; events beyond the agent's control can defeat assignments of responsibility. But this does not mean that we should forego consideration of responsibility for external harm. Conversely, the moral assessment of harm need not be defined in purely utilitarian terms. The difficulty with a utilitarian calculus is that even the projects to which individuals are committed as basic to their lives are somehow to be evaluated in terms of utility. But our moral projects are not so evaluated without loss of their motivating power. Smiley's point, again, is that we must link, but not conflate, causal and evaluative judgments in order to avoid rejecting the idea of responsibility due to the (causal) precariousness of human action or the assumption that the moral worth of our projects waits upon causal outcomes. She tries to show that by viewing responsibility as part of social practice we have the means to speak of accountability for external harm while recognizing the precariousness of human action.

Peter French, another social theorist of responsibility, has correctly argued that the primary motivation for a social theory of responsibility is the maintenance of identity within a community. This makes esteem and shame basic to responsibility. "Experiences of shame," he writes, "are characterized by a sensation of the loss or the slipping away of the identity one has tried to maintain and project to others. To be shamed is to be stripped of one's self-image."[25] The practice of praise and blame is not ultimately about guilt or innocence with respect to discrete actions as those actions are assessed according to rules or laws as in the case of agential theories. Smiley and French insist that one does not need to isolate a causal origin of every action to determine its moral status. For a social theory of responsibility, praise and blame function in terms of shame and esteem with respect to the social identity of agents and their conduct. What is at stake is the integrity of an agent's identity within collective life – how we bargain about responsibility and evade it – rather than the intentions of the agent or the utilitarian assessment of consequences of action. In so far as an agent's identity is at stake, the social practice of blame and praise is basic to a theory of responsibility.

Responsibility and Christian character

Marion Smiley presents a strong, social theory of responsibility. Social theories are found in Christian ethics as well. But these forms of thought are actually weak theories of responsibility because their primary concern is not responsibility assignments but the formation of Christian identity. A prominent exponent of a weak, social theory of responsibility in Christian ethics has been Stanley Hauerwas. We can isolate the salient features of this type of theory by briefly exploring his work.

Hauerwas insists that the first question of Christian ethics is not "what ought we to do?" but, rather, "what kind of people ought we to be in order to live out the story of God's action in Christ?" According to Hauerwas, Christians are to present an alternative vision of life to the violent forms of life found in the world, a vision of peace rooted in God's action in Christ.[26] The question of the meaning, formation, and moral implication of Christian identity is the central focus of his ethics. Yet the formation of Christian identity is not a matter of personal piety, an existential act of the will, or a refinement of some general religiosity, some ultimate concern, through explicitly Christian belief. Christian character is shaped and constituted through understanding life within the narrative of God's act in Christ as this story. is remembered and reinterpreted within the Church. The crucial social practice of the Church is telling the story of God's action in Christ.

For Hauerwas, thinking about human being and doing must take place within a community. Cognitive and moral judgments are deeply informed by traditions, tested in community, and embedded in practices of reasoning and moral formation along with the skills and virtues those practices entail. Moral rationality is tradition-constituted.[27] If we fail to attend to the import of communal memory, and specifically the Christian memory of God's action in Christ, we miss the grounds of valid moral claims. Thus, community and scripture relate dialectically in the formation of identity and in grounding Christian cognitive and moral claims. Scripture depends on a community for its interpretation; the identity of the community is constituted through telling the story of God's peaceable dealing with humanity in Jesus Christ. The "formation of texts as well as the canon requires the courage for a community consistently to remember and reinterpret its past."[28] Interpretation and remembering are political actions; they are about the life of some *polis*, some community.

Hauerwas counters those ethicists who seek to forge a universalist ethics "that does not depend on memory for its significance but turns to 'reason' and 'nature.'"[29] Like other social theories of responsibility, Hauerwas challenges any ethics, like Tillich's, which determines the norms for moral practice from a general picture of human existence, or, as Smiley puts it, from general metaphysical claims. Such an ethics frees itself from the Christian community and its memory of God's action. Hauerwas's point is that Christian existence is constituted by describing persons within the biblical narrative and understanding one's own life from that perspective. There is, we might guess, no Christian "self" outside of that practice.

The authority of scripture for a community "consists in its being used so that it helps to nurture and reform the community's self-identity as well as the personal character of its members."[30] Hauerwas follows the claim of Eric Auerbach that with the Bible "we are to fit our own life into its world, feel ourselves to be elements in its structure of universal history."[31] But Hauerwas wants to make even a stronger claim. The narrative also "creates more than a world; it shapes a community which is a bearer of that world."[32] Scripture does not only present a vision of the world, a cognitive and evaluative framework for understanding life, it also brings into being a community which can enact that vision in life. The moral task of the Christian life is then to appropriate and live out that story. Scripture and community are correlative realities; the moral life takes place within this identity constituting context.

The truth of the Christian life is bound to the life and practice of the Church. Truth is, in this sense, "like a 'knowing how' – a skill that can only be passed from master to apprentice."[33] The validity of that skill is confirmed by the community's ability to reconstruct its life relative to the narrative of God's action and different moral situations. The real problem for Hauerwas is the truthfulness of Christian existence with respect to the scriptural vision of life. Christian existence is truthful in so far as Christians understand and intend their lives *within* the narrative of God's action in Christ. The norm of such a life is peaceableness, a non-violent mode of life which enacts a distinctive possibility for human existence not found in the world. From this it follows that the primary Christian responsibility is not to reform the world; it is not to enact moral and political projects grounded in some supposedly general moral principle. The responsibility of the Church is simply to be the Church.

Hauerwas's account of the Christian life expresses a social theory of responsibility in that the question of the formation of moral identity through social practices is the focus of analysis. But for Hauerwas the founding social practice is not, as it was for Smiley, the practice of assigning praise and blame. It is, rather, the practice of forming Christian identity in truthfulness to the story of God's action. Reflection on the conditions for rightly assigning responsibility center on presuppositions for that social, interpretive practice. And for Hauerwas the crucial presupposition is not about human freedom or rationality; it is a claim about what God has done in Jesus Christ. A theological claim, not a claim about human action or norms for action, is the ground of the ethics. This is what specifies the distinctiveness of Christian ethics. It is also why Hauerwas's ethics is classified as a weak theory of responsibility. Neither human responsibility nor practices of praise and blame are morally central, rather God's action is the first principle of Christian ethics.

Social theories of responsibility

Social theories of responsibility focus attention on practices which constitute the identity and roles of persons and communities. The insight of this type of theory of responsibility is that the moral life is not simply about actions and choices; it is about the kind of people we ought to be. Action flows from character, and moral character is shaped and constituted through social practices and discourse which mediates those practices. In contrast to agential theories, this form of reflection insists on the irreducibility of the formation of character in the moral life. The self is an organ of the social whole; it is the community, not the individual, which is morally central.

Social theories rightly specify the fact that the integrity of human and social life is a function of the practice and discourse which form the identity of a life. We assign responsibility not simply on the basis of discrete acts, but also in relation to the character and social role or vocation of an agent. The very practice of assigning responsibility is one means, maybe the central means, of constituting moral identity. As we will see in Chapter 6, there is a reciprocal relation between identity and various practices of responsibility assignment. Ideas about moral freedom, about what defeats assignments of responsibility, and even about the scope of responsibility are a function of practices of forming and maintaining the identity of agents or a community of agents. In terms of an integrated theory of responsibility, social theories rightly examine the constitution and maintenance of moral identity with respect to social, linguistic practices. The integrity of an agent's life is in no small measure tied to the practices and discourse within which an agent's life is rendered intelligible. In Chapter 7 below, I develop the idea of radical interpretation to make sense of this point.

The difficulty in a social theory of responsibility, a difficulty Smiley recognizes, is to account for the authority of social practices. Why ought we to abide by the values and norms of our community? How do we determine the validity of those values and norms? The need to answer these questions was the driving concern of agential theories of responsibility, as we have seen. The conundrum, in other words, is how to account for both responsibility for self and the social mediation of moral identity. On reaching this problem, we can now turn to dialogical theories precisely because these theories take yet another approach to the problem of responsibility.

DIALOGICAL THEORIES OF RESPONSIBILITY

Dialogical theories of responsibility have enjoyed wide currency in modern theological ethics. They have taken two basic forms. One is seen in Karl Barth's divine command ethics; this theory has some continuity with agential theories of responsibility, but recasts the account of agency in terms of an encounter with the other. The second is the theory of responsibility developed by H. Richard Niebuhr as well as Roman Catholic moralists Bernard Häring, Charles Curran, and others. These theories have a marked social dimension, but, again, with respect to the demand to respond to others. The difference between dialogical theories is that for Niebuhr responsibility is the first principle of ethics, while for Barth the command of God is the answer to the problem of the human good. What an integrated ethics of responsibility draws from dialogical theories is an agentic-relational account of the nature of persons. Human beings are defined by their active relation to others, their world, and God. In this sense, an agentic-relational account of persons is also basic to the insights I will draw from agential and social theories of responsibility.

Responsibility and the command of God

Dialogical theories of responsibility center on the event of encounter between an agent and some "other." This "other" might be a human being or even the divine. According to these theorists, the self-disclosure of the other to us and how we respond to the other constitute us as persons. The encounter with the other shatters the dominance of the acting, knowing self in understanding and valuing the world and others. As Emmanuel Levinas has put it, the face of the other thwarts the drive to totality in which the meaning of reality is circumscribed within the view of the self.[34] The question then becomes, how is the event of encounter with the other understood within an ethics? Karl Barth's divine command ethics is one of the most radical expressions of responsibility ethics centered on encountering the other. He insisted, long before Levinas did, that the other, God, reveals himself and encounters the self as *commander*. While this would seem to be a "strong" theory, Barth's position is actually a weak, dialogical ethics of responsibility.

In answer to the question "what ought I to do?" divine command ethics asserts that in any specific situation one ought to obey the command of God.[35] This moral theory holds that the *meaning* of moral concepts – like good or duty or obligation – is equivalent to the idea of "commanded by God." In so arguing, divine command ethics stands in stark contrast to forms of ethics which ground morality in the self-legislating power of practical reason as Kant does, in general utility as utilitarianism does, or in moral customs and conventions as social theories of responsibility do. Divine command ethics is a normative moral position, but it is also a meta-ethical theory about the meaning of moral terms and the source of morality.

Karl Barth's ethics encompasses the normative and meta-ethical dimensions of divine command ethics. He explores the event of God being God for humanity as the necessary and sufficient backing of divine command ethics. The command of God confronts the moral agent in each and every moment of decision and thus in each and every situation in which the moral character of human life is determined. This is because, Barth writes, the "problem of ethics is the critical question in which the human sees his action placed, but that is simply his entire temporal existence."[36] To be sure, the Christian can place himself or herself in a better position to hear the command by attending to the preached Word of God. And there are also various spheres of life, as Barth calls them, which can guide moral reasoning. (These spheres are specified with respect to the action of God as creator, redeemer, and reconciler. For instance, the ordered relations in creation, the relation between man and woman being the most basic, are one sphere in which to hear the command of God.) Still, one acts rightly if and only if one acts in obedience to the divine command, the Word of God. As Barth writes:

When God confronts man with His command, what he wills is purely *ad hoc* actions and attitudes which can only be thought of as historically contingent even in their necessity, acts of obedience to be performed on the spot in a specific way, pure decisions the meaning of which is not open to discussion, because they do not point to a higher law, but is rather contained in the fact that God has decided in this way and spoken accordingly, so that human decisions can only obey or disobey the divine decision.[37]

The command of God does not point to a higher law or principle from which we could judge its validity. And given this, the only possible response to the command is obedience or disobedience. How we respond constitutes the rightness or wrongness of our existence.

In *Church Dogmatics* II/2, Barth claims that the ethical problem is the question of the good of human existence over and above all other goods. The problem is distinctively ethical in so far as for human beings to act is to exist. The good must be defined with respect to the problem of right human action. In making this claim about action and existence, Barth collapses the question of what is valuable for human life, that is, the good, and the question of what is morally right. Because for Barth to exist is to act, the rightness of human action is the same as the good of human existence. What then is the good of human existence?

The good of human existence is Jesus Christ, the Word of God, God's being-in-act. God's claim on human existence is grounded in God's self-giving to humanity and not in divine power. As Barth writes, "God gives Himself to man entirely in His revelation. But not in such a way as to give Himself a prisoner to man. He remains free, in operating, in giving Himself."[38] Barth further argues that by "deciding for God [the human] has definitely decided not to be obedient to power as power."[39] The meaning of God's freedom and power, the mystery of God, is manifest in God's being God for us, being gracious in Jesus Christ. Along with the equation of act and existence, this is the most crucial point in Barth's ethics. The being of God is understood with respect to the event of grace grounded in divine election, and not with respect to the exercise of power.

Knowledge of divine Lordship is constituted in the event of the revelation of God's grace, God's act which commands obedience. It is the actual Word of God, not the supposed possibility of divine power, that is the command of God. For Barth, we cannot reason behind the Word of God into the abyss of divine election. But it is not necessary to think behind the event of grace in order to ground the ethics. God is loving in the event of being God for us; that is the inner meaning and ground of the command of God. In Barth's terms, the Law is the form of the Gospel; the command of God is the form in which we hear and respond to the human good, the Word of God. Thus the meaning of the good, as a question in meta-ethics, and the actual command of God, the content of the normative ethics, are one and the same, that is, Jesus Christ. Any other definition of the human good is for Barth simply an expression of human sin, the human attempt to determine its good outside of, or other than, God.

For Barth a divine command ethics asserts the sovereignty of God in terms of the good and also in determinations of what we ought to do. What we think is good cannot be definitive of the good; what we decide to do is only right if it is in obedience to the divine command. This shatters our usual scheme of rendering life morally intelligible, and yet reconstitutes it through the Word of God. Since the Word of God has been incarnate in the true man Jesus, the moral life, while dependent on God, is preserved in its integrity as the answer to the human question implied in our capacity to act. The Word undercuts independent moral knowledge and thus all rational ethics outside of the divine command. Yet it also grounds an ethics of command in the being of God revealed in the Word, in Jesus Christ. The question of why God is worthy of our obedience is collapsed into the event of God being God for us and the command to obedience this event necessarily entails. Only in this event is the essence of God, the divine Lordship, knowable in terms of human existence or action.

Barth defines responsibility (*Verantwortung*) in terms of an obedient answer (*Antwort*) to the command of God. The Christian acknowledges, endorses, and conforms her or his life to the good who is Jesus Christ. Indeed, the basic command for Barth is really the permission to be obedient to the divine. Freedom for life, for others, and for obedience is the permission of the command of God, a permission to live our actual lives free from anxiety over the goodness of our existence.[40] Not surprisingly, Barth disallows the human response to the divine command to be definitive of the right. More properly, the human has already been obedient to God in Jesus Christ. In Christ, the right human response to the divine, the right obedience to the divine covenant, has been accomplished. The Christian's obedience to the command of God is at best an analogy of faith to Christ's action. What the responsible person does is to endorse Christ's obedient action. This ethics is then really, as Barth says, divine ethics; the focus of moral attention is the action of God and not human action. The question of the human good is taken into the event of God being good for human beings.

Responsibility as the principle of ethics

In contrast to Barth's divine command ethics, other Christian theologians have explored responsibility as the first principle of ethics. The self encounters the other not as a commander, but in patterns of responsive, social interaction. This is especially the case in the work of H. Richard Niebuhr. In *The Responsible Self*, Niebuhr develops his ethics based on the claim that "All life has the character of responsiveness."[41] Unlike Barth, this means that moral reflection cannot be developed simply from the dictates of dogmatics or the pages of the Bible. While informed by the biblical witness and Christian theological claims, the task of ethics is to aid us in understanding ourselves as moral agents.

In order to reach some level of self-understanding, ethical systems have used different symbols and concepts to apprehend the form of human life and to direct personal conduct. Niebuhr contends that three fundamental images or symbols have dominated Western accounts of the human as agent. The first account of what it means for us to be moral creatures focuses on the human capacity to determine which end(s) it ought to pursue and then to act towards that end(s). This capacity for purposeful action is what distinguishes persons from other creatures. "What is man like in all of his actions? The suggestion readily comes to him that he is like an artificer who constructs things according to an idea and for the sake of an end."[42] In Niebuhr's term, in this ethics the human is understood primarily as "man-the-maker." The dilemma of the moral life is what kind of life to fashion and what goods to seek. In terms of the argument of this book, the image of man-the-maker and the teleological forms of ethics which articulate it assert the moral demand to promote or enhance some end. Teleological forms of ethics differ in terms of which ends or goods ought to be pursued in human life. The second dominant picture of life arises from the fact that human beings are members in communities, we are citizens. For "man-the-citizen," the moral life is about rules for action and social existence. The moral problem of life is not simply which ends we ought to seek; it is how to govern ourselves and our communities in determining ends sought and means to those ends. This vision of the human finds expression in deontological ethics. Deontological forms of ethics differ in terms of the norm they articulate for life, the ground of that norm (for instance, is it in the will of God or human will?), how the law is known, and what makes it binding on agents.

Niebuhr sought to develop an agentic-relational account of the human. He focused on "man-the-answerer," the agent existing in complex relations of responsiveness to others. Niebuhr isolates the elements of responsibility as they constitute one complex, moral description of human life.

The idea or pattern of responsibility, then, may summarily and abstractly be defined as the idea of an agent's action as response to an action upon him in accordance with his interpretation of the latter action and with his expectation of response to his response; and all of this is in a continuing community of agents.[43]

In order to understand Niebuhr's ethics we can examine each element of this definition of responsibility.

First, responsible action is a response to action upon us. Responsibility begins with receptivity. Human life, Niebuhr avers, is known in terms of how a person responds to what he or she suffers and undergoes. Much as organisms interact with and respond to their environment, seeking to adapt in a fitting way to the larger whole which surrounds them, so too do human beings. We must understand that the self "is a being which comes to knowledge of itself in the presence of other selves and that its very nature is that of a being which lives in response to other selves."[44] What is required is then a fitting response to others.

Response to interpreted actions is the second element of responsible action. The distinctive character of human beings is that our responses to the world and others are not brute reflexes or determined by instinct. Were that the case, an agent could never be held morally accountable. All distinctly human responses are interpreted responses; they are mediated and directed by our understanding of the situation in which we live and also the meaning of actions upon us. As Niebuhr notes "we interpret the things that force themselves upon us as parts of wholes, as related and as symbolic of larger meanings."[45] Responsibility is responsive action in accordance with interpretations which seek an answer to the question or problem posed to us. In determining what we should do, we have to answer the prior question "what is going on?" Responsible action entails an answer to that question.

Thus far we have isolated two of the basic elements of responsibility: responsiveness and the response to interpreted actions upon us. The third element of responsibility according to Niebuhr is accountability. Our actions must not only respond to actions on us; we must anticipate responses to our actions in so far as we grasp the fact that all life has the character of responsiveness. Niebuhr's radical claim is that an agent is morally responsible for anticipated reactions to his or her actions. "Responsibility," Niebuhr writes, "lies in the agent who stays with his action, who accepts the consequences in the form of reactions and looks forward in a present deed to the continued interaction."[46] Thus, Niebuhr addresses the concerns of consequentialist ethics while also expanding the scope of responsibility to possible responses to our action.

Human actions have morally significant consequences. They must also be understood in terms of the context of continued social interaction. Social solidarity, as Niebuhr calls it, is the fourth element of responsibility. "Our action is responsible, it appears, when it is response to action upon us in a continuing discourse or interaction among beings forming a continuing society."[47] The self is born in the womb of society; our actions rely upon and contribute to the ongoing social reality of which we are part. We are citizens in some moral community and the moral problem is not simply the laws which govern that society, as in man-the-citizen forms of ethics; the basic moral problem is the scope of the moral community, the extent of our moral relations.

While all of human life is lived in particular communities with their own distinctive beliefs and traditions, the scope of the moral community, according to Niebuhr, is universal. This is because life is lived finally in response to the One God. This is the message of the Christian faith: the action we see patterned in Christ is "action which is fitted into the context of a universal, eternal, life-giving action by the One. It is infinitely responsible in an infinite universe to the hidden yet manifest principle of its being and its salvation."[48] Niebuhr's ethics insists on universalizability not with respect to the maxims of our actions, as was the case with Kant, but, rather, in terms of the moral community. In so far as this is the case, a strong, dialogical theory of responsibility involves beliefs about reality. This is found in the central claim of Niebuhr's ethics: "Responsibility affirms: 'God is acting in all actions upon you. So respond to all actions upon you as to respond to his action.'"[49]

Niebuhr's theory of responsibility, like Tillich's, articulates ontological claims about human existence and the world. This does not commit the so-called naturalistic fallacy of identifying the moral good as a predicate of objects like other natural predicates. Human agents must choose how to make a fitting response to others. And even the fitting response is not simply given in the situation of responsibility. Put in Niebuhr's terms, an ethics of responsibility articulates a relational theory of value. This theory holds that

> value is present wherever one existent being with capacities and potentialities confronts another existence that limits or completes or complements it. Thus, first of all, value is present objectively for an observer in the fittingness or unfittingness of being to being.[50]

In attempting to respond to others, an agent is then seeking a fitting response which complements or completes the other in its relation to God who is the center of value. Yet the domain of value "is not the multidimensionality of an abstract realm of essential values but rather the multidimensionality of beings in their relations to each other."[51] To understand all of life as a response to God is to know all things within the multidimensional domain of value. What is right, that is, a fitting response to others, is set within a relational theory of value, of goodness.

Dialogical theories of responsibility

We have been exploring dialogical theories of responsibility. These positions focus moral attention on the event of encounter and thus a relational account of the human being. Indeed, the wholeness, or, as I prefer, the integrity, of existence is constituted in the relations of responsiveness. Dialogical theories differ on this point, however. In Barth's weak, dialogical theory, the self is constituted by the other, that is, by the command of God who is Jesus Christ. This shifts moral reflection from the responsible self to the priority of the other, whether this be another person or the Word of God. Responsibility is not the first principle of this form of dialogical ethics; the self-disclosure, or revelation, of the other is the first principle of ethics. Revelation not responsibility is morally basic. This is why Barth must develop a divine command ethics. Niebuhr, as we have seen, claims that all life has the character of responsiveness. Yet since responsibility requires interpretation and accountability as well as responsiveness and social solidarity, the self is not constituted simply by the action of the other on the self. Responsibility, not revelation, is the first principle of ethics since the moral problem is how a self is to be responsible, that is, respond appropriately to others.

The insight of dialogical theories is to see that the integrity of human existence is intrinsically related to other persons, the surrounding world, and even the divine. In this respect, dialogical theories specify a basic claim of an integrated theory of responsibility in terms of an agentic-relational view of persons. But dialogical theories are unclear on the extent to which the drive to actualization, as Tillich called it, is valorized in the moral life from a Christian perspective. In so far as the self is constituted by its encounter with the other, what room is there for duties to self, for the demand to respect and enhance the integrity of one's own life as well as the lives of others? How is it that we are responsible for ourselves? Does responsibility mutilate human goods? On reaching this question, we can now draw some conclusions from our typological analysis of theories of responsibility.

TOWARD AN INTEGRATED THEORY OF RESPONSIBILITY

This typological analysis of theories of responsibility shows that theories of responsibility draw on diverse moral phenomena in speaking about responsibility. It is not at all clear that they are speaking about the same thing despite the use of a common discourse. We have also seen how theologians attempted to answer modern criticisms of mythico-agential views of reality in speaking about morality, criticisms we noted earlier in this book. They have done so through the connection between the will of God and our essential nature (Tillich), an appeal to the narrative formation of identity (Hauerwas), the event of the Word of God (Barth), or a specific theory of value (Niebuhr). In each case, a theory of responsibility was linked to a theological construal of the world in which human beings act.

We can now draw conclusions which specify the task facing an integrated theory of responsibility. First, an adequate theory of responsibility must account for the insights of these other positions about the nature of the moral life. This means, I judge, that we must conceive of the moral life as the dialectical relation between actualization of self and encounter with others mediated by social roles and vocations. To act morally requires that an agent understand the actualization of his or her own life in response to others with respect to the roles and vocations he or she has in social existence. This is the agentic-relational account of persons from the perspective of an integrated ethics of responsibility.

Second, an integrated theory must also account for the relation between the goods which persons are to respect and enhance and the principle for right action. This requires, I argue in the next chapter, specifying different levels of value rooted in basic human needs as well as articulating an imperative of responsibility. That imperative, furthermore, must mediate the self-understanding of agents who face the problem of the integrity of their own lives in response to others amid social roles and vocations. Only in this way, I judge, can we make sense of the demand to be responsible for self, the central insight of agential theories, while respecting the insights of dialogical and social theories of responsibility.

The problem of moral self-understanding leads to the final insight to be gleaned from our typological analysis of theories of responsibility. In different ways, each of these theories articulates a moral paradox. Agential theories articulate this paradox in terms of the form of life dedicated to acting on the moral law. In so acting a higher form of life, the good will, comes into being (Kant) or the multidimensional unity of life is actualized amid the travail of the moral life (Tillich). Dialogical theories understand the constitution of moral existence in terms of responsiveness to the other. By responding appropriately to others (Niebuhr) or in free obedience to the command of God (Barth), the self comes into being as responsible. Finally, social theories understand that moral identity is mediated by social practices of praise and blame (Smiley) or a community's narratives and interpretive practices (Hauerwas) with respect to roles and vocations. This means that to assume responsibility for our life is to have our identity constituted in terms of the goods and norms of our vocation. Moral identity is found through commitment to social roles. We gain ourselves morally by living responsibly within our callings.

An integrated theory of responsibility insists that a unique good arises by acting responsibly, the good of moral integrity. This is a unique good because it centers on the value of power to serve other values. An integrated theory must specify this insight about moral integrity so as to account for other theories of responsibility. And in so far as it offers a theological account of the moral life, it must do so with respect to the human relation to the divine. This final point remains open at this juncture in our inquiry. It will be redeemed only in Chapters 7–9.

CONCLUSION

The present chapter has provided a typology of theories of responsibility. By providing this typology, I have tried to isolate the insights and oversights of each type of theory. We have also been able to demarcate the range of topics an integrated theory of responsibility must address. The remainder of Part II of this book is dedicated to presenting an ethics of responsibility which can address these matters from a theological point of view. The validity of this ethics depends in part on the degree to which it can account for other theories of responsibility.

CHAPTER 5

Moral values and the imperative of responsibility

This chapter formulates the moral meaning of Christian faith in terms of the theory of value and the principle of choice of an integrated ethics of responsibility. Basic to this formulation is the idea of moral integrity. As John Kekes has noted, "Integrity is a complex notion. In one of its senses, it is principled action; in another, it is wholeness."[1] In this chapter, I am exploring, then, the wholeness of values basic to the moral life and also the principle for responsible action.

Christian ethics is committed to some form of realist moral theory because the reality of God, the ultimate human good, is prior to moral traditions or human invention. This realism is further elaborated in terms of the nature of created reality and what respects and enhances finite existence. I want to begin to develop the moral theory by considering the current debate about morality itself. This will enable me to clarify the kind of realism basic to Christian ethics, what I call hermeneutical realism. Later in the chapter I specify the principle of moral choice, that is, the imperative of responsibility, correlate to a theory of value. An integrated ethics of responsibility mixes the concerns of deontological and teleological theories within a multi-dimensional theory of value. This follows since Christians are commanded to love God and neighbor even as they are to seek first the reign of God.

THE DEBATE ABOUT MORALITY

In thinking about responsibility, contemporary ethics has taken it to be of utmost importance to determine whether persons and communities invent morality or discover it. Yet for most of Western history, ethics has been realistic in character. People believed that there was a moral order to reality, or, at least, to human life, and that the good life was about discovering that order and living in conformity to it. Stoic philosophers like Epictetus argued that there is a natural moral law written into the structure of reality. They believed that human beings can know this law through the use of reason and that this law should guide actions. Christian theologians, as noted in Chapter 3, insisted that God inscribed the moral law on the conscience. Conscience is a guide in human life because it grasps the distinction of good and evil, even if, they noted, we err in specific moral judgments. In spite of the differences between these ethical systems over the *content* of morality, the basic point was that to live a meaningful and praiseworthy life we must conform our lives to the way things are. The constraints on action and the possibilities for human life are not of our making; values and norms have roots deeper than the creativity of powerful individuals, the legacy of historical traditions, or the necessity of social life.

Traditional moral realism rested on a set of interlocking assumptions which have been rejected by the modern Western world. Most basically, classical realism claims that reality indicates what sorts of values we ought to seek. Disputes about what defines moral behavior are at root matters about the nature of reality and the nature of human existence. The advances of modern science and technology have thrown that conviction into question. Modern people do not assume that we can find out how we ought to live simply by looking at the natural world or natural tendencies of human life. Indeed, moderns find a good deal of their freedom in their capacities to escape these natural tendencies, for instance, through birth control. The idea that we invent morality arises from the criticism of the traditional belief that reality grounds morality.

In order to advance a plausible moral theory, we must examine the debate about realism in ethics. The basic argument for anti-realism, as J. L. Mackie puts it, is that "values are not objective, are not part of the fabric of the world."[2] In other words, the contention is that ethics is about *inventing* right and wrong. We do not *discover* moral truths, as traditional realism held, by examining human life, the nature of communities, or reality. Moral values are human constructions meant to serve specific purposes about social life, interpersonal relations, and individual well-being. The denial of the objectivity of values is not simply a linguistic claim about the meaning of moral terms; it is also an ontological claim. Values are not ontologically real.

Anti-realist arguments can take various forms. One form, called emotivism, foregoes the ontological point about values altogether, and, instead, develops a theory of the meaning of moral terms. Emotivist theories hold that when I say that you ought to be responsible, I only mean to commend that form of behavior. The status of moral claims is traced back, as it were, to the person uttering the claim. Moral norms and statements have no truth status beyond this emotivist meaning. Similarly, one could explore the social patterns of moral discourse in the attempt to grasp the meaning of responsibility. Social theories about the nature of moral claims deny the ontological status of values, but do not center on personal acts of commending values. They examine, as Marion Smiley does, social practices and a community's discourse of praise and blame. But that discourse and those practices have no backing beyond the community. The discourse of that community entails certain prescriptions for how persons ought to act, and also specifies the criteria for judging actions. However, Mackie is correct to argue that the most forceful form of all of these theories entails an ontological claim, that is, the denial that values are written into the fabric of the world.

Once the ethicist denies that values are ontologically real, the conclusion is that moral discourse is invented to serve certain personal and social needs. Morality involves a certain life project and has no backing beyond that fact. Anti-realist arguments about the nature of morality assert that our moral identities and norms are communal endowments which may or may not be expressed and validated in personal judgments. Just as values are not ontologically real with respect to the "world," they are also not real by virtue of something about human existence, say that we are rational animals, that we have certain desires, or are responsive. Yet, if that is the case, how does the anti-realist avoid a lapse into brute self-interest, the simple will-to-power? As I have asked before, how can the power to act be affirmed as a human good without itself defining the human good?

Anti-realists answer this question by appealing to the norms and values of communities which have shaped persons' moral identities. That is, they attempt to undercut the question by showing that there is no "self" who relentlessly seeks power in the service of her or his own self-interest; the very idea of a "self" depends on the moral vocabulary of some community. As Richard Rorty has put it, there is no human dignity that is not derivative from "the dignity of some specific community, and no appeal beyond the relative merits of various actual or proposed communities to impartial criteria which will help us weigh those merits."[3] The belief that somehow we are all at root Hobbesian egoists, or, as Christians believe, born into sin, assumes too much. These thinkers insist upon the intimate relation that exists between social existence, language, and the constitution of human identity. Identity and moral responsibility are without remainder the product of a social and linguistic community.

The point of this line of reasoning is to show that while persons often believe that their moral values have some hold in reality, are part of the fabric of the world, the "world" is actually a function of our language and social practices. Moral reason is bound to the discourse it learns and uses. The way to change our moral reasoning is to change our available discourse. Stripped of that discourse, we simply do not have any way to know what reality would be like, and we would not have any idea of who "we" are as persons and communities. Communities and traditions see and experience the world differently because of their moral outlooks and so we ought to explore those outlooks, rather than try to peer through them to "reality." As we saw in Chapter 1, the simple fact of the diversity of moral communities should chasten ontological claims for the status of all moral beliefs, including beliefs about responsibility. This is why Mackie rightly insists that his ontological claim is purely negative: values are *not* part of the fabric of the world.

This is an unacceptable conclusion for a moral realist. Realism designates the claim, or any theory used to explain the claim, that moral experience is not solely the construction of the conceiving subject or community. As David Brink has noted, in "moral deliberation and argument we try and hope to *arrive* at the correct answer, that is, at the answer that is correct prior to, and independently of, our coming upon it."[4] This has been a claim basic to most Christian ethics. And the reason is simple enough: for the Christian the reality and/or will of God, and not human needs or specific communities, are morally normative and exist prior to our understanding of them. Contemporary realists argue that if we are going to understand the tenacity of moral obligations and the sense of objectivity that surrounds moral experience, and not merely how we use moral terms, we must reject anti-realist theories. The ethical task is in fact to *discover* the truth about moral values.

There is an important assumption about moral reason which backs anti-realism in ethics. The assumption is, as Erazim Kohák notes, that "the way we perceive the world is a function of the way we conceive of it – and if the practice that follows from it is flawed, then the solution is to be sought in a more adequate conception."[5] But the moral theologian or philosopher can grant that all human understanding is dependent on some conceptual scheme or framework of belief and still hold that *perceiving* the good is morally central. It is this type of realism that seems basic to Christian ethics. The Christian is always trying to perceive the divine reality and purposes in the world. From the perspective of this form of realism, moral beliefs and cognitive schemes are empty without the experience that funds them. Ethics tests and revises moral concepts and beliefs by what are taken to be basic moral experiences. These experiences can be quite general in character, as, for instance, the experience of respect for persons. They can also be quite specific with regard to the moral situation of the individual or community. In each case, the task of ethics is to validate its claims by articulating the basic character of moral experience.

Different kinds of moral realism are possible. First, one can be a realist about a community's whole cognitive scheme or set of beliefs, since it is this scheme, and not specific ideas or experiences, that are about the world. Internal realism, as it is called, agrees with the anti-realist that people with different moral languages live in different moral worlds. "Internal realism," Hilary Putnam notes, "is, at bottom, just the insistence that realism is *not* incompatible with conceptual relativity."[6] This follows from the fact that to understand any specific moral concept, like responsibility, requires facility with the entire set of beliefs and concepts about reality in which that concept makes sense. For this kind of realism, moral reason is blind without some framework within which to understand and articulate our lives. One insists on realism in ethics with respect to the whole of a set of moral beliefs. The ultimate purpose of moral inquiry is to articulate and validate a claim about reality, but the ultimate truth of that claim is one towards which we are always struggling.

Theologians also make this argument, as we saw in Chapter 4. By insisting that Christian moral identity and understanding of the world are a function of the Christian narrative, thinkers like Stanley Hauerwas and John Howard Yoder mean to say that Christian beliefs are not definable in the terms of other forms of moral understanding. This saves the Christian moral vision from being reduced to claims about natural morality or the moral beliefs of the wider public. Yet it also means that the truth of Christian moral claims can only be established internally to those beliefs themselves. One cannot refer to the world "out there" to validate or to refute Christian moral beliefs. Those beliefs are validated with respect to communities whose life they help to constitute and whose life is a testimony to those beliefs. Oddly enough, this means that what becomes central is the question of the fidelity of Christians to the biblical narrative rather than the truth status of beliefs. Since truth is a statement about beliefs internal to a moral framework, the theologian explores the question of whether or not Christians are living truthfully by their convictions.

Other realists fasten on the question of validating claims about reality. They seek to establish, as Franklin I. Gamwell notes, "a logically necessary condition of human understanding by showing that its denial is self-contradictory."[7] This kind of realist hopes to show that the denial of some moral norm, concept, or position is

logically contradictory, and, therefore, the norm, concept, or
position is necessarily valid. Theologically this requires showing,
as Gamwell argues, that all persons have an implicit understanding of their existence with respect to the divine good.
Theism is then basic to ethics. Other realist thinkers adopt
phenomenological methods to validate moral realism. They
attempt to articulate the meaning of moral experience. In
Kohák's work, for instance, this means specifying a moral sense of
nature as constitutive of human existence. Still others, like the
Roman Catholic moral theologian Josef Fuchs, specify the experience of obligation in the phenomenon of conscience as definitive
for the meaning of being a person.[8] These stronger forms of
realism ground morality in the nature of existence, experience, or
self-understanding, rather than, as internal realists argue, the
coherence and comprehensiveness of evolving moral traditions.

Where anti-realists and realists differ is in how they seek to
validate their moral claims. The realist of whatever kind must try
to show how one position is rationally superior to another and
thus is true. He or she does so by trying to demonstrate a
position's ability to isolate and answer problems in other moral
outlooks. The realist insists that under *present* conditions it is
indeed difficult to refute the anti-realist, but holds, nevertheless,
that under some condition of knowledge, like (say) the eschaton,
that refutation will not only be a possibility, but in fact the case.
The anti-realist, conversely, can make no such argument. He or
she merely seeks to find points of commonality between moral
outlooks and to minimize moral conflict. But this does not mean
that one moral outlook is truer than another in any realistic sense.

I have been sorting my way through debates in ethics about the
nature of morality. Although Christian ethics has always endorsed
some form of realism, it has also placed qualifications on moral
realism. Historically, theologians argued that while the moral law
could be known without divine revelation, it is not to be equated
with the Christian moral vision. Most classical Christian theologians took the Noahic covenant and the Law given at Sinai to be
the sum of the moral law. Yet these are radicalized in the teaching
of Jesus, especially in the Sermon on the Mount (Matt. 5–7).
Because of the reality of sin and also the special content of Jesus'

double-love commandment, the Christian moral outlook was believed to correct, deepen, and radicalize natural moral knowledge. Christian ethics is thus realistic in that moral values are rooted in the nature of human beings as created by God, and, furthermore, the source of morality is God who exists prior to our understanding of the divine will and purposes. But this moral realism comes with the important proviso that moral knowledge requires interpretation not simply of "nature" or "natural laws," but also of the claims about Christ and human existence made by the Christian tradition. Through interpretation we discover the meaning of Christian faith for how to live.

From a theological perspective, the moral life is not simply a matter of creating or discovering norms and values. To insist that ethics is utterly creative is to overlook the simple fact that human beings depend on social, natural, historical, and biological conditions which foster and limit our creativity. We are created beings interdependent with the rest of reality. Yet to imagine that the human mind arrives at moral truths which somehow exist prior to our inquiry is to give a mistaken account of the experience of moral insight or spiritual discernment. Moral insight, the discovery of a moral truth about what we ought to do, involves, at least implicitly, some interpretation of a situation to which that "truth" is an answer. Interpretation is guided by orienting values and norms which are not simply read off a moral situation. In Christian ethics, we need an account of morality and moral agency which acknowledges creativity and discovery in how we arrive at judgments about what to do, and this account must correlate with an argument about the status of norms and values.

The route beyond the apparent impasse about inventing or discovering morality requires that we articulate the commitment basic to moral integrity and show its relation to the complexity of value. Further, this articulation must be correlated to an account of the nature of moral agents as self-interpreting, relational creatures, an account I develop below in Chapters 6–7. The moral project, I contend, is the commitment to integrate the goods of life such that human efficacy, the power to act, is put in the service of meaningful existence. An integrated theory of responsibility attempts, then, to render productive the conflict between realism and anti-realism in ethics. It asserts that we neither invent morality nor simply discover it. To amend a phrase from Paul Ricoeur, in ethics we invent in order to discover the truth of our moral condition.[9] We invent moral ideas based upon our experience of value, our perception of others and goodness, in order to discover our moral lot. I will call this hermeneutical realism in ethics. Like anti-realism and internal realism, this position grants that moral understanding is dependent on, and relative to, some cognitive framework. Yet with phenomenological forms of realism, I contend that perceiving, or experiencing, the good is morally basic. In terms borrowed from Charles Taylor and elaborated in Chapters 7–8, moral understanding is a matter of seeing-good and making-good. We invent in order to discover.

As the next step in providing a hermeneutically realistic account of morality, I want to clarify the enterprise of moral theory itself as one of the dimensions of ethics. Then I will outline a theory of value from the perspective of an integrated ethics of responsibility. This will be followed later in the chapter with the imperative of responsibility through which we are to interpret and guide our lives.

THE ENTERPRISE OF MORAL THEORY

Making a case for hermeneutical realism in Christian ethics requires that we be clear about two basic components of any moral theory. First, a moral theory must specify which values ought to be respected and/or promoted in life. Various theories of value are found in the history of ethics. A utilitarian, for instance, argues that we ought to promote the greatest happiness for the greatest number of sentient beings. What is of moral value, then, is happiness (however this comes to be defined). A Christian might be committed to the project of promoting the reign of God in human relations and the world, or, conversely, might understand the highest good in terms of the vision of God. In contemporary Western societies, as we saw in Chapter 1, the basic values are fulfillment and authenticity. But in each case, a moral theory advances a view about properties, states of affairs, or attitudes we ought to realize in the world.

It is important for my argument that a theory of value have two levels.[10] The first, lower level of value includes those goods judged to be worth directly seeking and promoting, like happiness, bodily integrity, or the social good. A moral agent has some sensibility for this level of goods since these goods, and their correlative disvalues, mark out the domain of human desire, needs, wants, and choices. While some moral theories do not grant it, there is also, second, a higher-level good. This good can be defined with respect to a commitment to respect and promote lower-level goods while itself constituting a unique good of meaningful life. As Bernard Williams notes, integrity involves an identification of a person with a "ground project or set of projects which are closely related to his existence and which to a significant degree give meaning to his life."[11] This higher good has to do then with the "ground project" in terms of which an agent or community of agents identifies the meaning of its life. To use theological language, this higher good is a matter of faith; it is fidelity to and trust in some project, be it, for instance, the good of a family, a profession, or a life in the arts, which defines the meaning and value of an agent's life. Admittedly, this is a unique form of the good because it concerns the projects to which agents are committed, projects which define the meaning and coherence of their lives.

This higher-level project cannot be defined simply in terms of lower-level values and sensibilities, and, given this, it cannot be respected and promoted in the same way as those values. It is, rather, a question of the commitments which define the agent's or community's life. At this higher level we are concerned with moral integrity in personal and social life. Not all theories of value recognize a second, higher level of value with respect to the identity of an agent. Because of this, the problem of moral integrity does not enter into those moral theories that deny a higher level of value. But Christian ethics makes this distinction between levels of value simply because it understands questions of faith to be basic to human existence. Faith is about what one trusts in and is loyal to in all actions and relations.

The first task of a moral theory is then to provide a theory of what ought to be sought in human life. Second, a moral theory must also specify a principle of choice in order to indicate which actions cohere with moral value(s). At issue in this second task is not what is valuable, but how agents should respond to values. This is sometimes called a theory of right correlated with the axiology, or theory of value, of an ethics. The principal division in ethics is between deontological and teleological theories. Each addresses the problem of choice differently.

A deontologist argues that there are some things which a moral person will absolutely never do. Let the right be done though the heavens fall, as it is sometimes put. The theory focuses on constraints upon human action and choice, the avoidance of wrongdoing, and thus the morality of the means in achieving goals. The deontologist holds, in other words, that "how one achieves one's goals has a moral significance which is not subsumed in the importance or magnitude of the goals."[12] To be sure, agents always act for ends; the deontologist's question is about the morality of the means used to reach them. And this is because respect for self and others takes priority over the goods which individuals may (or may not) pursue. One cannot violate the dignity or well-being of a person – oneself or another – in pursuing some end, no matter how praiseworthy, without morally demeaning that end. This is the deepest root of the contemporary value of authenticity; one ought never be untrue to oneself.

The teleologist argues that in any situation of choice one ought to promote values determined to be good, since it is the state of affairs realized by human action, including states of mind, that are objects of moral evaluation. From this perspective, the deontologist appears overly concerned with the authenticity of his or her own intentions, and, in fact, morally insensitive to the welfare of others. It is the case that the idea of what policy "promotes" the good and how one determines when the good has in fact been promoted or realized require further specification. But those are matters of calculation, not principle, and do not negate the theoretical point. As Charles Fried notes, "consequentialism subordinates the right to the good, while for deontology the two realms, while related, are distinct."[13] For the teleologist it is the ends and not the means which define morality. This finds expression in the contemporary value of fulfillment: any project worthy of devotion should fulfill and not mutilate the diversity of human goods.

Later in this chapter I will present the imperative of responsibility as a directive for choice. This imperative is concerned with how we ought to respond to values in actions and relations. The norm of choice from a Christian perspective entails both respecting and enhancing the integrity of life. It is a mixed theory of right. I want now to specify the theory of value for an integrated ethics of responsibility. Values rooted in the nature of human existence have realistic status even while the idea of moral integrity is unique in that it is not definable simply in terms of those natural goods. Integrity of life is a creative act, but in response to the multidimensionality of values grounded in basic human needs. This theory of value, we should note, is an answer to the problem of moral pluralism charted in Chapter 1 since it provides a framework for debating moral questions. The imperative of responsibility is to address the other pervasive problem of our time: how to exercise and direct human power.

THE DIVERSITY OF GOODS AND THE INTEGRITY OF LIFE

The moral domain of life is constituted by interlocking goods endemic to human existence and the choices we must make about them. Taken together these diverse goods constitute the values which the responsible person or community is committed to respect and enhance. This is to insist on a diversity of goods in the moral life and on the unity of life as a task. How might we think of the multidimensionality of value? I isolate pre-moral, social, and reflective goods. Moral integrity is what I will call the ethical good.[14] I want to outline these values here even though I will develop them in more detail over the next few chapters.

First, the moral life presupposes that human behavior is motivated by various biological, affective, and other vitalities, needs, and interests. These dimensions of life situate us in the world as feeling, aspiring, social, acting agents. The fulfillment of these needs is good; we naturally value things like sexual fulfillment, bodily comfort, food and shelter, and profound music. But this type of good is pre-moral in the sense that in certain circumstances it is permissible to violate or diminish these values while it is much more difficult, perhaps impossible, to justify acting directly against a moral value. For example, it is justifiable to sacrifice a limb to save a life even though this violates the pre-moral good of bodily integrity. Pre-moral goods place constraints on human choice, since choices are directed to preserving and enhancing these goods when possible. In this respect, moral understanding presupposes realistic claims about the world, human beings, and complex needs which form the background for human choices. All societies make claims about what I am calling pre-moral goods, even if they interpret and rank them differently. This is the point of descriptive relativism.

Second, the moral life entails standards, customs, rules, practices, and beliefs which communities develop to guide human life. These are admittedly human, social inventions aimed at respecting and enhancing values important for human well-being in relation to its social and natural environment, values like family, friendship, viable economic and political institutions, and means for interacting with the environment. These rules and norms for social action are the conditions for our thinking, speaking, and acting together. They provide a framework of meaning for the life of a community. By social goods I mean those forms of human excellence and well-being associated with fidelity to the consideration of others in a way of life and in specific choices. In so far as human beings must make choices about how to use, enjoy, and distribute pre-moral goods and also how to engage in social activities important for their life projects, social norms and goods are part of every life. One cannot act directly against these social goods without placing the very conditions of thinking, acting, and speaking at stake. Of course one can, and often must, challenge the prevailing norms as destructive, distorted, or unjust. But in doing so the protest itself must not act against the values those norms purport to protect: the conditions for our thinking, speaking, and acting together.

Finally, persons can and must assess moral beliefs, norms, and values and help to create new patterns of life. The critical analysis of conditions of human action and also of evolving moral beliefs contributes to life by enabling agents to be knowingly responsible for themselves and others. Reflective goods require the interpretation of beliefs and standards in order to reconstitute moral understanding and conduct with respect to a new insight into the moral condition. In terms I used before, this level of value expresses the range of human goods aimed at truthful life and self-understanding. These are the goods of culture or civilization, that is, the whole domain of symbolic, linguistic, and practical meaning-systems. Reflective goods entail a commitment to principles which protect and promote, rather than undermine, the meaning of a person's or community's life. What is entailed is a picture of human beings as self-interpreting agents who constitute coherent lives through judgments about what to do and to be in relation to others and the variety of goods which permeate their lives.

There is yet another level of good beyond pre-moral, social, and reflective goods. This is the ethical good. It is captured in the idea of moral integrity. Moral integrity characterizes the good of a life dedicated to respecting and enhancing the other goods of life in all actions and relations with all others. The meaning of integrity involves beliefs about persons, their social relations, and the meaning of ongoing, faithful action through time. As Lynne McFall has pointed out, "Integrity is a personal virtue granted with social strings attached. By definition it precludes 'expediency, artificiality, or shallowness of any kind.'"[15] "To have moral integrity, then, it is natural to suppose that one must have some lower-order moral commitments; that moral integrity adds a moral requirement to personal integrity."[16] While moral integrity cannot be understood solely in terms of lower-level goods, it is also unintelligible without them. Integrity specifies a particular attitude and project with respect to those goods. It designates a commitment through which an agent helps create her or his life with respect to goods at the root of personal and social existence.

Integrity specifies the wholeness of life with respect to some commitment. This is why we often think of the person of integrity as a man or woman of principle. But of course there can be serious and trivial commitments. We might admire the person who is committed to her stamp collection and still judge that this is not a morally serious matter. It is also the case that a person can be fanatically committed to some principle or way of life to the point of being destructive, hateful, or driven by vengeance. Thus while the idea of integrity is intrinsically related to commitments and projects, the question for ethics is which project *ought* to characterize a life. To put this again in theological terms, we all live by faith. The hard question to answer is the faith by which we ought to live. I return to this question below.

In order to clarify the distinctiveness of an integrated theory of value, it is important to note that the Roman Catholic philosopher John Finnis and theologians Germain Grisez and William E. May also insist on a variety of basic human goods. Grisez, for instance, argues that there are reflexive goods which involve making choices (self-integration; practical reasonableness; friendship and justice; religion or holiness), substantive goods which stand apart from the idea of choice (life and bodily well-being; knowledge of truth and beauty; and skilled performance and play), and complex goods which have reflexive and substantive aspects (marriage and family). An adequate ethics, Grisez insists, must acknowledge these kinds of goods. What is more, Grisez, Finnis, and others argue that since these are different kinds of good we cannot develop a concept of the good to commensurate the variety of goods. If that is true, then we can never act directly against the realization of any human good, a claim Grisez, Finnis, and May argue other moral theologians erroneously insist upon.[17] But this also means that the principle of moral choice cannot be derived from these goods even if it ought to protect them. The basic principle of morality, as Grisez formulates it, is then: "In voluntarily acting for human goods and avoiding what is opposed to them, one ought to choose and otherwise will those and only those possibilities whose willing is compatible with a will towards integral human fulfillment."[18]

The integrity of life, I have argued, is its wholeness, or fulfillment, in terms of the multidimensionality of pre-moral, social, and reflective goods. Despite an insistence similar to Grisez, Finnis, and others on the variety of basic human goods, an integrated ethics of responsibility offers a distinctive theory of value. Moral integrity demands truthfulness *of* self *to* the project of respecting and enhancing the integrity of all life. The idea of moral integrity relates to and yet transforms the values of fulfillment and authenticity found in contemporary life. Moral integrity is an absolute value. Actions and relations, and the intentions and motives they entail, which denigrate and destroy the integrity of life or evaluate the worth of others simply in reference to oneself or some social purpose are always immoral. However, we are concerned with the integrity of the whole of life and not simply the sum of its constitutive goods in isolation from one another. And this means, against Grisez, Finnis, and others, that in some circumstances we are justified in acting against certain goods, like pre-moral goods, in the name of the whole of life. The moral life is simply ambiguous.

What moralists call the principle of double effect admits of the ambiguous character of human life. Namely, one ought always to intend to respect and enhance the integrity of life, realizing that in so acting some unintended disvalues might follow from our action. An agent ought not directly intend a moral evil, even if in acting some disvalue might occur as an unintended consequence of the action. This means that moral reasoning demands analysis of the disvalues which will result from an intended action with respect to the integrity of a person's life or the common good. The reality of moral ambiguity, rather than some abstract idea of the good which commensurates all values, is why an integrated ethics of responsibility requires agents to make tough, even tragic, decisions between competing values. In this respect, the theory of value developed above diverges from the arguments of Grisez, Finnis, and others who hope to avoid a tragic dimension to the moral life.

Further, the point of the imperative of responsibility as I develop it below is not only, as Grisez and Finnis argue, to guide human choices. The imperative enables us to formulate a judgment about the truth of our moral condition and with this paradoxically to constitute a new mode of life in the world. The imperative expresses the integrity of life before God and specifies this as a principle of choice. The basic claim of Christian faith is that the integrity of human, historical existence is an indirect consequence of a commitment to the project of seeking justice, loving mercy, and walking humbly with God (Micah 6:8). Integrity is achieved not by directly aiming at it, but with respect to the dictates of morality. As Hans Jonas has noted, the "secret or paradox of morality is that the self forgets itself over the pursuit of the object, so that a higher self (which indeed is also a good in itself) might come into being."[19] Moral integrity is the concept Christian ethics must use to specify this "higher self" which comes into being, is enacted, through a commitment to respect and enhance the diversity of goods which characterize life. This self "appears" in answering the claim of these values uttered in the existence of others. We have a sense of values ranging from simple pleasure to an awareness of the moral worth of other persons that thwarts the ego. The moral paradox is that integrity is not achieved in acts of self-creation so much as it is graciously received by seeking to live by the imperative of responsibility. By living responsibly, we gain ourselves. The Christian lives by faith in a double sense: one lives in loyalty to a specific project of respecting and enhancing the integrity of life before God, and, furthermore, one lives in trust that this commitment is constitutive of the human good.

An integrated ethics of responsibility charts the diversity of goods and insists that moral integrity is not reducible to nor separable from them. Morality in this sense is not simply invented; it entails a construal of human existence aimed at making judgments about how to live given these pervasive features of life. Morality is also not simply discovered written into the fabric of the world since we must interpret features of life in order to understand them morally and to make right choices. Every interpretive act and each choice involves the creative use of reason. This account of morality makes sense of the ideals of fulfillment and authenticity but transforms them in terms of moral integrity. Moral integrity is about the fulfillment of life but with respect to a commitment *of* self *to* be true to a specific moral project. Again, I will develop this theory of value in more detail over the following chapters. Now I want to turn to the principle of choice for an integrated ethics of responsibility. It is meant to specify the project or faith which ought to guide our lives.

MORAL INTEGRITY AND THE IMPERATIVE OF
RESPONSIBILITY

A moral imperative is a principle of choice about how to live in all our actions and relations. It specifies the standard for determining whether an action, relation, intention, or choice is moral or immoral. In the theory of value presented above, a unique good, the good of moral integrity, arises out of acting on the imperative of responsibility.

In order to specify the imperative of responsibility we must first draw a distinction between prudential and moral reasons for acting. This distinction will help us to understand in what respect the imperative of responsibility is *binding* on all our actions and relations. Consider an example of what seems to be a moral statement: "In all our actions we ought to respect and enhance the integrity of life in others and in ourselves." The statement is a *practical* claim since it concerns human action and is a directive for action; it also seems to be a *moral* claim since it specifies how we ought to live with respect to values. Yet I can understand the meaning of the statement and not act on it. I can admire persons dedicated to the responsible life and yet decide against so dedicating myself. In so far as the statement includes a principle of choice, I can always decide not to act on that principle. But the force of the statement is that in so choosing, I am acting immorally. I am violating a principle which ought to be acted on in all my actions and relations. The above statement is not simply reporting or describing something about the human condition. What then is the character of our knowledge of those practical and moral claims?

Beliefs about which values we ought to seek presuppose as the condition for their intelligibility claims about agents, as agential and dialogical theories of responsibility have rightly insisted. This means that any statement about what to do must presuppose and also bear upon the coherence of the life of the agent who is trying to act on that principle. Without some degree of coherence no conceivable agent could act because there would, in fact, be no identifiable agent. But practical statements are directed to agents. Thus every statement of this sort must presuppose and contribute to the coherence or fragmentation of an agent's life. The importance of this fact ought not to be missed. It means that the idea of the integrity of an agent is logically and ontologically prior to the goods which he or she can or ought to seek; it means that an act of commitment to live with some integrity is prior to the quest to secure certain values in existence. Moreover, it also implies that the agent who is choosing and acting has unique status with respect to those other goods. It means, in other words, that we can ask about the integrity of a moral agent, and this is precisely what the above statement does in formulating a principle of choice.

If we have settled the fact that a *practical* statement is directed to some agent and that this logically entails a question about the coherence of the life of that agent, we have still not differentiated a *moral* imperative from other possible maxims of action. Perhaps the statement is not moral in the strict sense but *prudential* in character? If the statement is in fact a prudential one, then it must be read as follows: if you want to live responsibly, then you ought to respect and enhance the integrity of life. In this form the imperative would be, as Kant called it, hypothetical in character. It would specify the means which ought to be used to reach ends which are desired or wanted. The application of the norm would be contingent upon the agent(s) seeking some end or good which the norm will lead to attaining. A moral imperative, conversely, concerns the ends as well as the means we ought to respect and seek in all actions and relations. It obtains in all cases, even cases where our desires and wants are lacking, where there is weakness of will.

Directives for action based on the idea of integrity are good candidates for categorical moral imperatives. This follows since an imperative about integrity is directed to agents and the meaning of the lives of these agents depends on the project(s) to which they are committed. The identity of an agent – including his or her loves, wants, volitions, and needs – is bound to the projects which direct and give coherence to that person's life. Again, without some degree of coherence no conceivable agent could act, because there would be no identifiable agent. The idea of integrity articulates the relation between identity-conferring projects and the acting person. Thus an imperative which concerns the integrity of life must be categorical since its object is a necessary condition for other imperatives and choices of whatever sort. In this sense the imperative is a requirement binding on the exercise of human power. It shows us what necessarily impinges on all moral experience and deliberation. I explore the theological source of this claim below in Chapters 7–8.

An imperative about moral integrity is distinctly moral in character and yet this does not mean that it denies prudential motives in human action. Persons seek coherence in their lives; that is, I argued in Chapter 2, a basic human aspiration. There are prudential reasons for following the imperative of responsibility. However, the imperative is moral since it specifies what kind of coherence we ought to seek in life. Moral integrity denotes a way of life in which certain actions cannot be done if the person or community are to survive as they are. As Fried notes:

> what constitutes doing wrong to another may also be regarded as the denial owed another's moral personality ... All other values gather their moral force as they determine choice. By contrast, the value of personhood ... far from being chosen, is the presupposition and substrate of the very concept of choice.[20]

There are some things a person must never do simply because such action would undermine personal life. That which directly and intentionally destroys or demeans the meaningful coherence of diverse goods and aspirations is categorically prohibited. And this follows because the idea of integrity is basic to the idea of the agent as the one to whom the imperative is directed.

The imperative of responsibility is this: *in all actions and relations we are to respect and enhance the integrity of life before God.* Each term in the imperative and the relations among them are important. The fact that the integrity of existence, and not responsibility, is the heart of this moral outlook indicates that the idea of responsibility plays a different role in this ethics than in some forms of Christian ethics. It means that we are concerned with the coherence of the goods of life along with the question of the project to which we should be committed.

An imperative for how to live responsibly is directed to agents and thus necessarily focuses on *actions* and *relations*. This is because human life is a matter of activity and we live within complex sets of relations with others; we are active, social, and responsive beings. We exist within the changing, pulsing, living ecosphere. "Life includes the totality of earthly existence with its intricacies and lushness, its raw pain and barrenness, its solitude and interrelatedness, its movement and multiplicity, its resoluteness, its violence and terror, its harshness, its surprises."[21] Human beings manifest the dynamic of life in our existence as agents related to others and the world. The multidimensional theory of value offered above specifies this fact at the level of good. We must decide how to live as persons in relation with others in so far as we meet situations in which it is not clear what we are to be and to do. The imperative of responsibility is to guide our decisions about how to respond to the domain of values.

How ought we to live as agents? The imperative of responsibility centers on *respecting* and *enhancing* the integrity of life. In our actions and relations we first have the obligation to *respect* the integrity of life. This means that the lives of others and ourselves make a claim upon us to recognition and regard in all actions, a claim I explicate more fully in Chapters 7–8. In so far as we are to respect the integrity of life, the lives of others or our own lives cannot be treated merely as means to other ends. Respecting the integrity of life provides content to what it means to respond rightly and justly to others. One is to avoid harm, oppose cruelty, and also to combat forms of humiliation which demean the integrity of the lives of persons and communities by destroying their frameworks of meaning. The demand for respect has deontological standing in terms of the levels of value rooted in creation. Life makes a claim on us to be respected. This claim to respect is a manifestation of the source of morality, the divine. The realism of Christian ethics is rooted in a perception of the divine as the source of morality revealed in finite existence. Yet Christians are also enabled and required to *enhance* the integrity of life in all actions and relations. This means working for the flourishing of life in its several dimensions. Not only are there constraints on how one ought to act and relate to others and ourselves which are grounded in respect, there is also the positive injunction to labor for the well-being of life in its fullness, its integrity. Theologically, this means seeking to enhance the reign of God in all actions and existence. Respecting and enhancing the integrity of life is the *content* of what it means to respond appropriately to others, the world, and God.

The ordering of these injunctions within the imperative as well as their mutual implication is extremely important for Christian ethics. To reverse the order of the injunctions would be to invite the judgment that actions aimed at the possible integrity of some life override the demand to recognize and respect existing forms of life. But we ought not to seek to *enhance* the lives of others and ourselves if so acting would violate the demand to *respect* the integrity of life. Respect forms a moral baseline, as it were, for all conduct and forms of social life. This is because the giving and withdrawing of respect constitutes the scope of the moral community. In all actions and relations there is an overriding demand to acknowledge others as part of the moral community and thus to acknowledge their claim to respect. This is grounded in God's claim on human existence to recognition and regard for the divine being as good. As we have it in the first commandment: I am the Lord your God, you shall have no other gods before me. (cf. Exod. 20:2–3; Exod. 20:23; Deut. 5:7). Recognition and regard for the other, especially the divine Other, constitutes the sphere of moral existence. The demand to enhance life is situated within a duty to respect the integrity of life. The ordering of the injunctions to respect and enhance the integrity of life is to insure that the power to affect the world, and thus to enhance values, is always in the service of goods other than itself. The categorical force of this imperative derives, then, from the fact that "integrity" denotes both a necessary condition for the intelligibility of any moral claim (respect), since it is directed to agents, and also a task to be undertaken (enhance).

Yet we cannot assume that respect alone is what persons are morally required and empowered to be and to do. Theologians have long held that Christian faith entails the belief that we are to labor for the flourishing of life within the context of a fundamental respect for the integrity of existence.[22] Doing so requires, of course, moral imagination and discernment, since what will enhance flourishing cannot be determined a priori. The task of enhancing the integrity of life means seeking to maximize pre-moral, social, and reflective goods. For instance, a Christian has responsibility for seeking to relieve human suffering but also to provide the means for education and the development of persons' mental, creative, and cultural capabilities. Similarly, we are to respond to social values by working for the just treatment of all persons in collective life. And one must also seek to enhance the integrity of non-human life.

This does not mean, as critics of consequentialism insist, that one is driven to the conclusion that there is one single end to seek in all action, nor does it place on agents an oppressive burden to do good. The reasons for this are simple. The imperative directs us to increase the diversity of values and thus to contribute to the richness of life. What is more, our moral projects are always limited by the demand to respect the integrity of life in others and ourselves. Monotony and fanaticism are ruled out of the moral life. The point I am making is that the deontological dimension of an integrated ethics of responsibility finds its correlate in a teleological vision of well-being. The imperative articulates a mixed theory of right. It relates the insights of deontological and teleological theories by charting their difference but mutual entailment.

The imperative is directed to acting, social agents. It requires that we respect and enhance the integrity of life. Living responsibly means respecting and enhancing the *integrity of life*. The most basic problem in ethics is to determine what defines the integrity of life. This question is at once empirical and ontological. Any answer requires claims about what integrity is and also about the concrete, actual conditions which contribute to or destroy the integrity of life. By integrity of life I mean, first, the integration of the vitalities, needs, and interests of life into some coherent and identifiable whole. All forms of life actively integrate the vitalities and processes of existence into complex constellations of relations which constitute the being of that form of life. This is true of molecular processes in organisms and also, albeit in different ways, of societies. In so far as any entity or person exists, it is characterized by some degree of integrity, some coherence in its existence through time in relation to others and its environment. It has the power to be, as thinkers like Paul Tillich put it. Without some degree of integrity of existence through time, an entity would simply cease to exist at all. In distinctively human action and relations, an agent or community of agents struggles to integrate the goods rooted in the bodily, social, and reflective dimensions of existence and the needs, vitalities, and interests these entail.

However, we are also self-interpreting agents. Persons feel, sense, and also want to understand the finite, fallible, and fragmentary character of their lives and their communities. Our lives manifest a restless quest for coherence. This relentless quest is the origin of ethics. Human life is always lived with commitments which are to bring coherence to existence. A person's or community's identity-conferring commitments, in Jeffrey Blustein's apt words, "reflect what they take to be most important and so determine to a large extent their identities."[23] Accordingly, the idea of integrity designates, second, how life ought to be lived. It specifies a uniquely moral project which arises out of and in response to the fact that we sense the finitude, fallibility, and fragmentary character of our lives and yet seek wholeness. The imperative of responsibility specifies an identity-conferring commitment.

All life despite its fragmentariness is characterized by some measure of integrity; that is what it means to exist. The moral question is what kind of integrity ought to characterize our lives in so far as through the power to act we can make decisions about what way of life to live. This "ought" is not reducible to the fact of existence, since we can integrate our lives in many ways. It is possible to live with personal integrity and yet be morally misshapen. One can imagine the dutiful Nazi who fulfilled his terrible role in the execution of Jews. Moral integrity cannot be simply identified with the fact that all human beings require some degree of coherence with respect to identity-conferring commitments. This is why moral integrity is not definable in terms of authenticity or truthfulness to self. Moral integrity entails the commitment to a specific project objective to self within which the agent's identity is constituted. And yet the moral "ought" is also unintelligible without reference to the dynamics of existence. What one seeks is coherence with respect to the complexity of value in life.

Moral integrity designates the integration of the goods of life with respect to attitudes and commitments to a moral project which defines what an agent's life is about. It is a norm not reducible to the brute fact of existence, but which concerns consideration of the well-being of others as well as self. Commitment to responsible living is consistent with the fact that to exist is to have some degree of integrity but also to seek fulfillment under the demand of respecting and enhancing life. Moral integrity is the ethical answer to the problem life poses to us of fashioning healthy, just, and coherent personal and social lives. For instance, if the bodily integrity of life is not respected and enhanced, it is impossible for life to continue. But we have to make judgments about how to respond to bodily existence. By the same token, if social relations are not respected and enhanced through political, economic, and cultural means, the human project falters and fails. And this requires specifying how the cohesiveness of social life ought to be justly lived. Finally, in so far as human beings can reflect on the meaning of bodily and social existence, the quest for meaning and thus for the truth about our lives must be respected and enhanced. To suppress or distort the human quest for meaning and truth destroys not only social relations but personal existence as well. Human life is characterized by the struggle to integrate truthfully and meaningfully bodily, social, and reflective dimensions of existence. This struggle arises out of our inarticulate sense of the finitude, fallibility, and fragmentary character of human life.

The imperative of responsibility means that in all actions and relations we ought to respect and enhance the dimensions of value and their integration into a coherent and truthful way of life. However, what we are responding to within the moral life is not simply a diversity of goods or even the way persons and communities happen to work out a way of living. We are to respect and enhance the integrity of existence manifest in personal, social, and planetary life. The idea of integrity articulates not simply a fact of life as it is, but the irreducible worth of life to which we ought to respond, a worth manifest, albeit in fragmentary ways, in our actions and relations with others. Moral understanding and relations are thus mediated through the idea of integrity which articulates the worth of existence. The moral problem in large measure is how to see others as fellow strugglers besieged by life and seeking wholeness. This, again, is the realism of Christian ethics; it is the attempt to see the creator in the creature. We interpret our lives and world morally through the imperative of responsibility in order to develop our ability to see and understand the worth of existence and how best to respect and enhance the integrity of life. That is the moral point of what I have called hermeneutical realism.

Traditional moralists would have spoken of this formation of moral perception by some ideal or concept of goodness in terms of loving others and ourselves *in* God, loving and respecting persons as the image of God, or, in Kantian ethics, respecting *humanity* in persons. Because actual life is finite and fragmentary and our moral sensibilities are fallible, we are to respect and enhance the integrity of life ambiguously present in our own lives and the lives of others by acting on principle. For the most part, we do not find others making an unambiguous and immediate moral claim on us. It is a fact of human existence that we turn our eyes and close our hearts to others and even ourselves, out of fear, perplexity, revulsion, or apathy. This is the problem of moral callousness, and, for Christians, the moral reality of sin. In this light, one of the major errors of contemporary ethics of responsibility based on the call–response paradigm is its unending search for moral immediacy; the quest for some unmediated experience of the moral worth of others or the command of God. Any honest assessment of our moral condition must note how self-deceptive such arguments can be and also how fleeting these experiences of the "other" are in human life. This is why the formation of moral sensibility is crucial in an ethics of responsibility: we must learn to see the irreducible worth and dignity of life through the fragmentariness of life. The formation of sensibility, I argue in Chapter 7, is bound to a certain identity-constituting act called radical interpretation. The point here is that respect for and enhancement of life is always mediated by some idea, symbol, event, or principle of what is ultimately worthy about persons and communities. I am to respect and enhance the lives of others and my own life not because any specific life is actually characterized by integrity and thus somehow merits moral attention, but, rather, because this idea enables me to grasp the truth of existence despite its ambiguity. This is what Christian faith means by knowing and loving others *in* God.

An ethics must consider, then, that in relation to which the worth of life is found. Non-theistic forms of ethics contend that the integrity of life can be defined with respect to our social lives, rooted in the forms of life we fashion as individuals, or grounded in our interactions with the natural world. The values we ought to secure in existence are thereby understood with respect to a community, the self, or the natural environment. These social, existential, and naturalistic ideas of what grounds the worth of life have dominated much modern and contemporary ethics. Christian theological ethics contends that the ultimate integrity of life is found in relation to that which transcends our communities, our lives as individuals, and even nature itself. God is that in relation to which the integrity of all of life is found, a relation which appears fragmentarily in the goods which characterize finite life. Moral integrity is found through faith in the God of Jesus Christ and therefore the relation to the divine is constitutive of existence. We are, of course, to respect and enhance life in terms of the worth of society, personal life, and brute existence. But persons and communities are also to respect and enhance the transcendent worth of life endowed by the living God. I continue this argument in Chapter 8 below.

This returns us to the fact that we are speaking of a moral imperative. In so far as the norm which is to guide our actions and our relations is formulated as an imperative, we are required and empowered honestly to face and actively to change the ways in which we demean and destroy the integrity of life, our own and that of others. The imperative of responsibility is a critical principle for judging current ways of life and social relations. It specifies not only how we ought to live, but also what forms of life ought never to be lived. It distinguishes moral from immoral actions and relations; the imperative helps us to expose personal and social fault and sin. By doing so, the imperative exposes the false and destructive ways that life is integrated, valued, and used by unjust works of power. Put differently, the imperative of responsibility is a principle for orienting how we should live because it defines the moral domain of existence. It is an answer to the prevasive problem of how to direct human power.

The imperative of responsibility formulates distinctive features of Christian faith in contemporary terms. It articulates the prophetic injunction that we are to seek justice, love mercy, and walk humbly before God. Justice and mercy find expression in the demands to respect and enhance the integrity of life, demands which can only be fulfilled in humility before God. To speak of the integrity of personal and social life in relation to the divine is to insist that moral goodness is a matter of right relations to God and others. In Christian faith, moral integrity, or right relations in personal and social life, is the form the good takes in history. Thus, the idea of integrity conveys in contemporary terms the biblical concept of righteousness as the sum total of goodness. Our lives are morally right and good in so far as they are integrated through faith in the God of righteousness and active in the respect and enhancement of life. Communities are just when they provide the means to respect and enhance the integrity of persons and the common good. Respecting and enhancing the integrity of life before God is the defining mark of the responsible life. The imperative of responsibility articulates these themes of prophetic Christian faith.

CONCLUSION

In this chapter I have developed an account of the diversity of goods and moral integrity and outlined an imperative of responsibility from a theological perspective. This theory transforms widely held beliefs about fulfillment and authenticity and provides a principle of moral choice. It answers the problem of moral pluralism in that certain goods are basic to human life even while, descriptively speaking, cultures interpret these differently. Agreement about the levels of goods provides a context for moral debate and judgments across cultures. The imperative of responsibility is formulated to clarify Christian convictions and also to answer the pressing problem of human power in our time. The next task, then, is to explore in more detail the one to whom imperatives are addressed, the moral agent. In considering that question, I turn from the problem of morality, the subject of the previous chapters of Part II of the book, to the other major theme of responsibility, the nature of moral agents. In doing so, I will bring greater specification to the theory of value and the imperative of responsibility outlined in this chapter.

CHAPTER 6

Freedom and responsibility

Every ethics entails some theory of the nature of moral agents. For an ethics of responsibility, a moral agent is understood in terms of patterns of responsibility assignment and the freedom, or power, to act in relation to others and the world. The present chapter examines the complex relation between moral freedom and patterns of assigning responsibility. Peter French notes that "we seldom ask if someone was acting freely unless we are concerned with pinning responsibility for something on him or her." Questions about responsibility "provide the conceptual superstructure in which questions of metaphysical freedom or moral freedom are conceived."[1] Yet it is equally clear that in order to make sense of assignments of responsibility, and thus praise and blame, we must establish human efficacy, the power to act. The purpose of this chapter, then, is to chart the dialectical relation between freedom and responsibility assignments. The sections of this chapter move between these topics. In the next chapter I explore responsibility and moral identity. Taken together, these two chapters present the theory of moral agency of an integrated ethics of responsibility by articulating an agentic-relational view of human beings.

THE QUESTION OF AGENT RESPONSIBILITY

Earlier in this book, I drew a distinction between causal and agent responsibility. The distinction turned on the degree to which we could in a stream of events identify an agent worthy of praise or blame. The question of agent responsibility is not simply "what caused this to happen?" but, rather, "who is responsible?"

The identity of an agent, "who" one is, is bound to the task of responding to, and answering for, actions and relations with others and to oneself. As J. R. Lucas notes:

the central core of the concept of responsibility is that I can be asked the question "Why did you do it?" and be obliged to give an answer. And often this is quite unproblematic. But sometimes I cannot answer, cannot be expected to answer, the question "Why did you do it?", and then I say "I am not responsible."[2]

Agent responsibility is about the connection between (1) an agent as a cause in the world and (2) how that agent is or is not held responsible for actions. In trying to rid himself or herself of responsibility, a person might show that he or she did not do the act, was mistaken for someone who did, or was coerced into acting.

In order to grasp this point fully, let me recall examples of dilemmas of responsibility assignment. In the creation story Adam attempts to shift the burden of the responsibility for his sin to Eve by claiming that she initiated the act ("She gave me to eat"). The history of interpretation, tragically, has seen this as an act of Eve coercing or seducing Adam to act. Adam appeals to Eve's suggestion as the reason for defeating God's assignment of responsibility to him. This is his sole means of ridding himself of responsibility, since, given the nature of the divine command, he clearly could not plead ignorance. Similarly, the gangbanger, to recall another example, might plead that he was not the one who fired the gun which killed a child, but was mistaken for the person who did. Conversely, his attorney might insist that the gang member was coerced into acting or was ignorant of what he was doing. An abusive parent might plead insanity about an act of sexual abuse, claiming that he or she simply lacked the cognitive capacities required to be a fully responsible human being.

In each case, we speak of the person as the origin of deeds for which she or he must give an account. If one cannot give an account of one's actions as one's own, then one's moral agency is in question. The capacity to give an account, to answer the question "why did you do it?", is intrinsic to the meaning of being an agent, a person. When I say, "I did that", who I am is manifest in a particular situation in relation to what or who questions me about some action. Yet assuming responsibility, as well as the legitimacy of punishment for misdeeds, rests on clarity about the extent to which an agent meets the conditions necessary for an action to be voluntary. If an action assigned to me does not meet those demands, then either it was coerced, it was an involuntary reaction, or there is a conflict between what I am willing to appropriate for myself and what others want to ascribe to me.

The rightness of assignments of responsibility turns on the question of voluntary action. We are responsible for our actions and to others because in some basic sense we possess, or own, our actions and our lives. An agent is self-directed or autonomous if she or he exercises some measure of control over her or his life. And this is actually the force of the claim, made by classical authors like Aristotle, Augustine, and Aquinas, that the principle of voluntary action must be *internal* to the agent. An ethics of responsibility requires what Wolfgang Huber has called the reflexive use of principles by agents with respect to the demands of a moral situation.[3] In deciding what to do, the agent brings moral principles into action and in doing so helps to constitute her or his own identity within the social duties she or he bears. This is an act of freedom *internal* to the agent.

At this point we need to explore further moral freedom as the necessary condition for agent responsibility and hence the assignment of responsibility. I want first to isolate different types of theories of freedom. Then I will explore the freedom required for responsible action. Throughout these sections of the chapter I will also examine the extreme case of responsibility assignment, that is, punishment. The discussion of freedom and punishment will allow us to complete the dialectical inquiry of this chapter by considering patterns of responsibility assignment even as it will also raise issues to be addressed in the next chapter.

RESPONSIBILITY AND TYPES OF FREEDOM

In his famous essay "Two Concepts of Liberty," Isaiah Berlin distinguished between negative and positive senses of liberty or freedom. Negative freedom "is the arena within which the subject

– a person or group of persons – is or should be left to do or be what he wants to do or be, without interference by other persons." Positive freedom, Berlin notes, concerns "the source of control or interference, that can determine someone to do, or be, one thing rather than another."[4] While Berlin's distinction is drawn with respect to political freedom rather than the moral life, it is important to explore these senses of freedom since they have shaped ideas about responsibility and punishment.

The idea of negative freedom has been important in modern liberal political and moral theory concerned to preserve a space of freedom untrammeled by the demands of the State or the Church. The most basic moral principle is that we are free to undertake any endeavor as long as it does not harm another person's capacity to pursue whatever she or he desires. This conception of freedom does not deny the fact that persons may have to give up some freedom or constrict the domain of personal rights in the interest of social stability and the protection of rights deemed important. Liberal theorists from Thomas Hobbes to John Rawls have insisted on this fact.[5] The point is that the theory of negative freedom entails the moral principle of not harming others, or of respect for personal rights. The theory offers no substantive account of value, no prescriptions for the kind of life persons *ought* to seek.[6] The assumption is that it is possible to conceive of freedom without articulating its connection to some idea of the good. We ought to pursue our personal wants, seek self-fulfillment, and be true to that project, be authentic. These are, in fact, the basic values which dominate late-modern societies, as I charted earlier in this book.

A negative conception of freedom is concerned with the arena of liberty. What destroys freedom and thus renders an agent no longer responsible for his or her conduct is simply unjustified interference. Responsible conduct requires the limitation of coercion in the social sphere. This is because, as Berlin writes, if "I am prevented by other persons from doing what I want I am to that degree unfree; and if the area within which I can do what I want is contracted by other men beyond a certain minimum, I can be described as being coerced, or, it may be, enslaved."[7] A theory of negative freedom holds that what renders an agent no longer accountable for her or his deeds is unwarranted coercion. An agent is responsible if and only if he or she acts freely, where acting freely is defined negatively.

Now, if we link this understanding of freedom with the claim to neutrality about what good we personally ought to seek, then we can understand the peculiar idea of responsibility found among some liberal theorists. John Stuart Mill put it concisely. Mill argued that "responsibility means punishment."[8] When we ask if we are responsible for some action, we are really asking, Mill reasons, whether or not we can be rightly punished for it. Responsibility is tied to social practices which center on the legitimate constriction of the domain of liberty (punishment), because the only moral claim on the individual is to respect others' rights to noninterference in the pursuit of their wants. One is not responsible for the kind of person one is, since, as shown, that argument would require some substantive idea of the human good. Responsibility means punishment because on this account of freedom only actions which wrongly interfere with others' lives can be judged morally, that is, be open to punishment. Interestingly, this means that the larger the arena within which I can freely pursue my wants without interference with others, the less I am ever capable of being responsible, being punished. Responsibility is contrasted with freedom negatively conceived; my consideration of the rights of others and the possibility of my being punished stand against the domain of my freedom.

Here is the difficulty with the idea that "responsibility means punishment." If there is, first, no connection between responsibility and some conception of the good, and, second, if responsibility means punishment, then, third, on this theory there is actually no way to specify the purpose of punishment, or, what is the same thing, of responsibility. This follows because without a conception of the good we cannot make sense of punishment serving any moral purpose, but, per definition, punishment, as inflicting what is unwelcome on an agent, must serve some moral purpose if it is to be just. What is the purpose of being responsible? It is unclear in the terms of a negative conception of freedom why one ought to be responsible outside of fear and/or

self-interest, which is to say there is little reason for being moral –
as opposed to prudential – on this concept of responsibility. Once
the contrast between this conception of freedom and the idea of
responsibility is grasped, it is hardly surprising to find criticisms of
responsibility in the name of personal fulfillment and authenticity
in societies dominated by classical liberalism. The conclusion
implied in this line of reasoning is that somehow human life is
qualitatively better the less responsible we must be. This brings us
to a positive conception of freedom.

What Berlin calls the positive sense of freedom means that per
definition a moral agent is a creature who does have some control
over his or her life.

> The "positive" sense of the word "liberty" derives from the wish on the
> part of the individual to be his own master … I wish to be somebody,
> not nobody; a doer – deciding, not being decided for, self-directed and
> not acted upon by external nature or by other men as if I were a thing,
> or an animal, or a slave incapable of playing a human role, that is, of
> conceiving goals and policies of my own and realizing them.[9]

Since freedom is not defined simply in terms of non-interference
with one's wants and projects, this means that liberty has some
essential connection to a substantive moral vision. To be free is to
be in control of our lives and to realize purposes. However, to be
self-directed entails some conception of what one ought to seek
even if this is just greater self-direction. Freedom is a matter of
being able to enact an ideal or normative vision in actual life.

The fact that freedom is linked in this positive conception to
some moral ideal means that we must draw a distinction between
what a person is or wants to be and what an individual *ought* to be
and to do in order to enact his or her true humanity. The
purpose of the moral life is to realize the higher self, a claim I
have made with respect to the idea of moral integrity. On this
account, an individual could be free from external coercion and
still not be genuinely free or self-directed. As Charles Taylor has
noted, "Doctrines of positive freedom are concerned with a view
of freedom which involves essentially the exercising of control
over one's life. On this view, one is free only to the extent that
one has effectively determined oneself and the shape of one's

life."[10] In contrast to theories of negative freedom, a positive theory of freedom insists on the essential connection between freedom and responsibility. Contrary to what proponents of negative freedom argue, we can be unfree in ways far more subtle than the limitation of our pursuits by others. This means, as Taylor further writes, that "you are not free if you are motivated, through fear, inauthentically internalized standards, or false consciousness, to thwart your self-realization."[11] This claim about the range of attitudes, beliefs, and dispositions which can render an agent unfree is unintelligible from the perspective of a negative theory of freedom.

It is clear that a positive theory of freedom does not claim that "responsibility means punishment." Yet a positive theory warrants a theory of punishment. In the name of the "higher" self and for the sake of reforming the offender, it can warrant "the deliberate infliction of something unwelcome on a rational agent."[12] Positive theories of freedom are committed to justifying punishment not simply in terms of prevention or deterrence, but, rather, for purposes of reforming the wrongdoer, or, in retributive theories, to vindicate the law and the victim. The correlation between freedom and a vision of the human good thus exposes the main danger in positive theories of freedom. A positive conception of freedom can lead to coercion when those in power believe their idea of the "true self" ought to be forced on others for the sake of the good of those others – the forms of moral tyranny exercised by Church, State, and traditional moral authorities against which liberal theorists reacted. In the history of Christian ethics, positions have ranged from the execution of heretics in order to save them from further endangering their souls through sinning to an insistence on the freedom of conscience and religious practice from State intervention. The question a positive theory of freedom raises is thus about the limits to promoting a substantive vision of the good. Critics of positive theories of freedom rightly ask: what are the limits to aiding others in their quest for the good life? In terms of this book, there are certain baseline requirements of respect for others which circumscribe the arena of promoting values and purposes based on a vision of the human good. The demand for respect curtails the extent to which we may and must interfere in the lives of others under the requirement of enhancing the integrity of life.

With Berlin's help, we have isolated two types of theories of freedom and how they relate to responsibility and punishment or, more generally, what one can be blamed for. An integrated theory of responsibility must account for the insights of these theories of freedom without purchasing their faults. I have already suggested that it does so in part by specifying the demand both to respect and to enhance the integrity of life. In terms of a theory of punishment, this means relating deterrence and reformatory practices aimed at the integrity of the community and the criminal's own life. That said, we must deepen our discussion of freedom and responsibility by further considering the very capacity to act and thus to be responsible for actions and outcomes.

THE NATURE OF MORAL FREEDOM

A basic assumption of ethics is that we must be able to speak of human efficacy. If we ask about the minimal conditions for that efficacy, we can say that in order to be morally responsible a person must have the "ability to do otherwise." If an agent is not able to do otherwise, due to compulsion, violence, or ignorance, then she or he is not acting voluntarily. In so far as this is true, it is difficult to hold that agent morally responsible. Of course, given certain beliefs an agent can be responsible even when he or she was not able to do otherwise. In Christian ethics, thinkers like Augustine, Calvin, and others have argued that agents are responsible for their character and conduct even though they were not able to act otherwise than in sin. These claims are strictly theological in character; they concern the status of the agent *before God*. And given this, only God can redeem fallen, sinful humanity; human beings cannot merit grace through their actions. Yet this argument does not obviate the *moral* insight. In order to be held responsible an agent must act voluntarily even if human action is circumscribed within a religious vision of reality. All that is required from the perspective of traditional Christian theology, as Augustine noted, is that one's will be in one's power where "in one's power" means that one acts on an internal principle.

The problem for ethics is that the phrase "able to do otherwise" is ambiguous. It can be taken in at least two distinct ways, which lead to different theories of the nature of moral freedom and responsibility.[13] Different problems confront responsibility ethics depending on how we understand the simple phrase "able to do otherwise." In order to grasp this point let us return to the example of Adam and Eve in the Garden of Eden.

First, this phrase can mean that an agent must be able to act on a principle of choice, and, furthermore, that this choice must not be determined by the situation to which an agent responds or by her or his inclinations and interests. I will call this the "voluntarist" model of freedom. It holds that we would not consider Adam and Eve responsible for their act of sin if we learned that God had implanted some device in their brains that made them desire the apple. Likewise, if God foreknew that Adam and Eve would sin, then they were also determined to act come what may and thus were not responsible for their action. The reason that Adam and Eve are not free under these conditions, the voluntarist argues, is that given these particular conditions one can explain motives, intentions, and decisions without reference to Adam and Eve as the origin of their choices. The voluntarist insists that for an action to be free any account of it must have reference to the acting agent in such a way that the description does not picture the agent as yet another causally determined event in the world. The problem with deterministic metaphysics, whether theistic or not, is that the relevant description of reality under which we are to understand the act of eating the apple is not Adam's and Eve's own sense of being agents who choose to eat the apple, but, rather, God's omniscience, or, in secular arguments, a causally determined universe. If we can validate a description of reality that is deterministic, what becomes of Adam's and Eve's responsibility? The voluntarist answers by saying that these persons were not responsible if they were not free to do otherwise than they were determined to act.

The claim of this position is then that an agent is free and responsible if and only if the grounds of the agent's choice are constituted in the very act of choosing, or the self legislates its own maxims of action. Moral responsibility is born of the capacity to refer to ourselves as the origin of certain actions which we "own," and which, for good or ill, affect the world. When I say "I did that," my identity is specified as the point of origin of a particular act which I confess to own. Being responsible is simply acknowledging that we were the origin of our own deed. On this line of reasoning, a theory of responsibility must explain the connection between agent and deed, and, what is more, the way in which the agent is a force in the world subject to moral evaluation. This is the concern of what I called in Chapter 4 "agential" theories of responsibility and why they focus on the agent/act relation.

We can isolate a conundrum in the "voluntarist" model of the ability to do otherwise rooted in the fact that human beings are defined by relations to others and the world. Since every agent acts in some world, a description of the world must, at least implicitly, factor in our conception of the agent. We cannot conceive of a human being otherwise than in a world, and what we mean by a "world" is always peopled by agents. From what perspective ought we to provide a description of the world, and, assuming we can answer that question, is such a description in fact possible? The problem is that in explaining the world in which an agent acts, we seem to deny the voluntarist model of freedom to do otherwise. This is the case since the *explanation* of an "event" requires the ability to specify the "causes" of the event within a sequence of events, and also to predict, within limits, possible future courses of events.[14] In so far as this explanatory task is at all possible, the category of "action" is better redefined as *event*. The determinist dispenses with the idea of action in her or his account of the world. The world is a constellation of "events." It is this claim which led some modern thinkers, like Kant, to defend a dualistic conception of reality, that is, the claim that reality is composed of a phenomenal realm of events governed by strict causal laws and a noumenal realm of freedom under moral law. Simply put, the voluntarist interpretation of the "ability to do otherwise" means that freedom and determinism are incompatible. Ethics must be independent of claims about reality. This is the root of anti-realism in ethics, as we saw in Chapter 5.

The insight of the voluntarist model for interpreting the ability to do otherwise is that it makes sense of our experience of human efficacy. When I exert myself by acting in the world, I believe, quite rightly, that it is I and not someone or something else who acted. The meaning and value of my life is tied to the fact that I am able so to act, to be a doer and decision-maker rather than an object acted upon. I know myself not simply as a part in some causal chain of events, but *as* an agent capable of intervening and directing the course of events and my life. The connection between an agent and her or his acts is important not only for the moral assessment of action, but for human self-understanding. This is why, as I noted in Chapter 1, some strategies of ridding oneself of responsibility are dehumanizing; they entail the claim that one is a product of determining forces and thus not an agent in one's own right. In addition, this is the reason why modern thought tends to picture the world as having only one kind of agent, the human agent. In order to be free, an agent must be the sole cause of an action and not determined by the world round about her or him. A disenchanted picture of the world is crucial to this theory of freedom.

For these reasons a further difficulty becomes clear about a voluntarist theory of the nature of moral freedom. The difficulty centers on the connection between my "true" or "higher" self, manifest in acts of freedom, and the person I am in the world with others and subject to desires, wants, and the ravages of time. Can I make sense of who I am outside of some consideration of the commitments, values, ideals, and beliefs I hold about what is good and true? Is the "true" self, as existentialists argue, reducible to a principle of pure choice, a being stripped of relations, commitments, and desires that simply chooses? Are we really most free when we exist outside of relations with others, isolated in a domain of pure, unfettered choice? Must we forego a robust sense of ourselves in order to preserve moral freedom?

The phrase "free to do otherwise" is open to another inter-pretation. We can understand this phrase with respect to human wants. On this account of the ability to do otherwise, which I call the "evaluative model," it is not the capacity for choice but the evaluation and formation of wants that is central. We need only to specify the kind of freedom required for moral self-understanding and need not decide whether freedom is the constitutive condition of (noumenal) reality or the transcendental ego. As Gary Watson writes, the "problem of free action arises because what one desires may not be what one values, and what one values may not be what one is finally moved to get."[15] The basic contention for this interpretation of the ability to do otherwise is that human beings have distinct and diverse sources of motivation, and, what is more, these sources can conflict. I might want to stop some troubling habit, say smoking, at the same time I desire the "good" that habit entails (for example, the satisfaction of physical and mental desire). Our lives are riddled with these kinds of conflict. An evaluative theory argues that an agent is free if and only if she or he acts on what is most basically valued, what really matters to her or him, and not simply what is desired or wanted. I am not really free if desire overrides my considered commitment to quit smoking. This is what traditional moralists have called the weakness or bondage of the will, a concept, we should note, unintelligible on the voluntarist model where the only possibility is free (i.e., authentic) action or unfree conduct. The fact that what we value might be shaped by social roles, conventional beliefs, natural desires, and needs does not negate moral freedom if we come to endorse those values.

This line of reasoning about freedom is long-standing in Western ethics. For instance, Plato spoke of the desire of reason as the desire for the "good" which is a distinct source of motivation from other human desires, desires usually associated with the body and irrational passions. St. Paul bemoaned that what he knew he ought to do was not what he did; the "flesh" was at war with the "spirit" as a distinct source of motivation. Augustine spoke about the conflict of love of self and love of God in the moral life. He insisted that a sinner exercises free choice even when his or her will is bound to a sinful love of self and the hatred of God. This follows since the sinner endorses as the principle of his or her action the love of self. Indeed, that endorsement is the meaning of sin. Moral identity is constituted around multiple sources of motivation and the kinds of persons we choose to be with respect to them. A person comes to know himself or herself in terms of these often conflicting sources of action and the choices he or she makes about them. Human inwardness and self-understanding are born within this struggle at the heart of existence. I expand on this claim in Chapter 7 by isolating basic sources of human action.

How does this relate to the problem of freedom and responsibility? The fact that human agents have some form of self-understanding is logically entailed in the idea of a moral agent. However, the actual self-understanding of any particular agent is always a contingent and not a necessary fact. My moral identity is not determined by my being an agent. In principle we can change or revise our self-understanding and thus our identities. And it is this fact which is basic to what the "evaluative" account of the ability to do otherwise means by moral freedom. The freedom to revise one's life through the examination of values and desires is not incompatible with determining factors on our lives. Of course, it is the case that we cannot utterly change ourselves. And change itself is often a slow and torturous process, as anyone knows who has tried to change his or her "habits." This is why Christian theologians have argued that a radical change in the being of the human is beyond the power of persons; it is a matter of divine grace. Granting the difficulty of change in the moral life, persons can interpret their lives and make revisions in their self-understanding. This process is basic to how we live in the world as moral beings and indicates the meaning of moral freedom.

This exposes the problem in the "evaluative model." If freedom is always tied to desires and wants, how is it that we know this fact except by the conflict between those desires and wants and some other principle of action which, in certain moments, we act upon? Knowing that our freedom is bound by desires and wants means that we sometimes act through the effort of free will, we act on the claims of conscience, which exposes to us our wants and desires. This is what Kenneth E. Kirk once called "the argument from the unknowability of the constant factor." Whenever "anything is constantly present to an observer from the dawn of life to its sunset, it must remain eternally unknown." More precisely, if "there is anything which, in so far as it enters the sphere of human experience, is absolutely constant, continuous and universal, it is something of which the human mind can have no experience."[16] But if that is so, then the ability to do otherwise cannot be totally explained in terms of the evaluative model for interpreting freedom. It is this insight that has driven the voluntarist model of interpreting the ability to do otherwise.

We have seen that the way freedom is understood shapes an account of responsibility and the validity of acts of praise and blame, including punishment. The two models for interpreting the freedom to do otherwise entail conundrums which seemingly cannot be answered in the terms of the models themselves. I have also tried to show how the positive and negative theories of freedom must complement each other in order to avoid coercion or a loss of the grounds for morally evaluating acts of punishment. In the next chapter I will develop a modified positive, evaluative theory of moral freedom and responsibility. It is modified by the fact that a Christian acts on a principle other than simple wants and volitions, specifically through faith in God. With some sense of the debate about moral freedom as the condition of responsibility, it is now possible to return to the question of the assignment of responsibility. What are the patterns of responsibility assignment and what defeats these assignments? How are these tied to ideas of freedom?

THE ASSIGNMENT OF RESPONSIBILITY

Several questions must be considered in order rightly to assign responsibility to agents: (1) who is responsible? (2) to whom is an agent responsible? (3) for what is an agent responsible? and (4) what defeats assignments of responsibility, or how are excusing factors defined? It is also the case that the assignment of responsibility takes two distinct forms. We *ascribe* responsibility to others and thus hold them responsible; we *assume* responsibility for ourselves and thereby appropriate responsibility for our character and conduct. Let us explore the two forms of responsibility assignment, mindful of the range of questions which must be answered in order rightly to hold others or ourselves responsible.

Distinct concerns come to focus depending on whether we are ascribing or assuming responsibility. Indeed, these activities seem to demarcate different "territories," to use Albert Jonsen's term, in moral reflection.[17] If we ask about assuming responsibility, we unavoidably must consider the identity of the acting agent and her or his commitments, a sense of self. Conversely, if we hope to ascribe responsibility to others, then a host of concerns about social roles, motives, and grounds for excuse will enter consideration. The perplexing question is the extent to which these two activities overlap. Does one form of assigning responsibility depend for its validity on the other, for example, does ascribing responsibility require that agents first assume responsibility for themselves? These are, of course, long debated questions, and the various theories of responsibility differ in how to answer them. But the point of contact between these activities is that in each case the individual or community assigning responsibility is attempting to render the power to act subject to standards of evaluation. This is what makes acts of assigning responsibility themselves matters of responsibility from the perspective of an integrated ethics. We can succeed or fail morally in terms of how we assign responsibility. In so far as one rightly or wrongly ascribes or assumes responsibility, one helps to bring about a moral event – some contribution to or detraction from the human good – simply because in this act the agent is contributing to or detracting from the good.[18] Anyone who has been falsely accused, wrongly held responsible, understands how the ascription of responsibility can detract from the human good. The proper ascription or assumption of responsibility contributes to the good by affirming the reality of agents.

This is the reason why the assignment of responsibility is linked to questions of punishment. The legitimacy of any act of punishment rests on the extent to which it contributes to the good. In order for it to do so, responsibility must be rightly assigned; errors in assigning responsibility detract from the human good because they destroy or deform personal or communal life. Theories which legitimate punishment in terms of deterrence or prevention seek to render the guilty agent unable to detract from the human good. Theories which focus on the reform of the criminal are geared towards making the agent take responsibility for her or his actions and thus to direct future action. Vindictive theories of punishment, primarily interested in paying back those wronged or upholding the law, seek to affirm for the victims and the rest of the community the demand that agents are obliged to contribute to the human good and not detract from it. Against Mill, punishment, I judge, must be seen in the service of responsibility and this means that it ought to respect and enhance the integrity of life through deterrence and reforming the criminal. The propriety of actually punishing an agent, or, conversely, rewarding an agent, requires then clarity about the assignment of responsibility.

In order to consider these matters, we can explore in detail some cases of assuming and ascribing responsibility. The paradigm of *assuming* responsibility for some action or character trait can be formulated as follows. In response to the question "why did you do it?" an agent asked to assume responsibility (1) confesses having done the action ("I did it"), her or his intention to do it ("I intend to do it"), or denies having done or intended the action ("I would never think of doing such a thing … "), and (2) when pressed can in principle provide some justificatory reason for or account of the action ("I did it because … ", "I want to do it because … ") or excusing reason ("I didn't do it because I was in California at the time"). In terms of the *ascription* of responsibility, the person, institution, or community seeking to hold an agent responsible must determine (1) that the agent did the action in question or intended to do the action, and (2) that there were no excusing reasons for the action or that the agent acted involuntarily. If we recall, again, the case of Adam and Eve, we see these features of our paradigms.

In ascribing responsibility to Adam for eating the apple and then punishing Adam and Eve, God seeks to determine that Adam in fact ate the fruit of the Tree of the Knowledge of Good and Evil. According to Genesis, God makes a determination with respect to Adam's newly found self-knowledge, that is, knowledge of his nakedness. God also administers punishment with respect to factors which bear on reasons for action. Adam, Eve, and the

Serpent are all punished differently with respect to their actions in bringing about this detraction of the good. In being questioned by God about his nakedness, that is, in being asked "did you do it?", Adam tries first to deny that he did the act, and, then, when asked for reasons for his conduct, tries to shift blame to Eve. The propriety of God's ascription of responsibility to Adam could be defeated by showing that Adam did not do this act – the act was either not done at all (the apple fell from the tree) or was done by someone else (say, Eve). It could also be defeated if certain excusing factors were present. Why are Adam and Eve culpable if God knew they would eat the apple? Divine foreknowledge could be a reason to defeat the assignment of responsibility to Adam and Eve.

The paradigm and the biblical example help to isolate several factors noted at the beginning of this section of the chapter important for understanding the assignment of responsibility. First, *ascriptions* of responsibility must meet a demand of a factual nature. In ascribing responsibility to an agent a causal judgment must be made about whether, in fact, the agent or community of agents in question acted and whether the action was actually done. This means that the ascription of responsibility tends to focus on the act and the *consequences* of an action, rather than possible intentions or proposed actions and the character of the agent. This is not necessarily the case with the *assumption* of responsibility. An individual can take on the responsibility of another person to the point of vicarious suffering whether or not he or she actually did the action. This capacity to "take on" the responsibility of another is fundamental to what I called in Chapter 3 "representative action." An individual can decide to represent others and thus assume their responsibility. The assumption of responsibility does not face the same demand for a causal judgment as the ascription of responsibility. All that is required for the assumption of responsibility is that whatever or whomever is holding someone responsible be willing to accept the transfer of responsibility to another. This has been a crucial point in traditional Christian theories of atonement. Theologians from Athanasius through Anselm and beyond insisted that Christ became "man" because only in his doing so can the guilt of sin be rightly transferred to him and thus satisfy divine justice. As Athanasius put it, "what he has not assumed he cannot save." Ascriptions of responsibility, on the contrary, must meet certain factual demands; they entail strict causal judgments.

Generally speaking, ascriptions of responsibility are *ex post facto*; they are formulated retroactively. As J. R. Lucas has noted, in the justification of an action, and, we can say, in the ascription of responsibility, there is a "double temporal standpoint, analogous to that of the perfect tense."[19] In ascribing responsibility to someone now for some good act, we are attempting to say what happened and why it happened. Generally, we do not ascribe responsibility for intended actions or for mental states, motives, or purposes. It may be the case that in certain societies there are designated "thought police," or, as in totalitarian states, the demand for (the appearance) of ideological conformity. As Vaclav Havel has noted, in communist nations persons were expected to live in untruth, not to admit their true intentions and purposes since the State claimed a right to that sphere of life. Granting this, ascribing responsibility is, in general, retroactive and has the linguistic form of the perfect tense.

In the Christian tradition it is possible to speak of prospective ascriptions of responsibility and also of ascribing responsibility to agents for their mental states, purposes, and motives. God judges the "inner man," as it was traditionally put. In so far as God searches the human heart, the divine makes ascriptions of responsibility in ways usually inaccessible to human judgment. This has meant two things in the history of Christian ethics. First, it has meant that certain dimensions of human existence are free from the scrutiny and control of secular authority. The State or society cannot command conscience, as Martin Luther put it.[20] One cannot be commanded by the State or society to believe certain things, to forsake religious conviction, or betray one's own conscience. Roman Catholic ethics asserts that an agent must always act according to conscience, realizing, of course, that conscience can err in many ways. Persons are free from the ascriptions of responsibility by external authorities in certain and specific senses. Conscience alone, as the internal voice of God, can judge purposes, thoughts, and motives. Second, in the history of Christian ethics the claim that only God can judge the human heart has also resulted in a growth in human inwardness and scrupulosity, as I detailed in Chapter 1. Christian theological ethics has been deeply concerned with the spiritual condition of agents, the orientation of our loves, the purity of intentions, and the quality of faith. An act viewed "externally" might be morally neutral or even praiseworthy, an instance of what Luther called civil righteousness, and still be blameworthy before God because of the agent's motives. The extent to which radical circumspection about our motives and purposes before God contributes to excessive guilt is a debated matter in current thought, as noted before.

Nevertheless, ascriptions of responsibility generally depend on determining matters of fact about what was done and who did it. Ascribing responsibility entails a causal judgment: it involves determining who acted, whether the act was in fact done, and, most importantly, whether or not the act and agent "caused" a moral event. The second thing the paradigm case reveals about ascribing responsibility is the question of whether or not an action is in fact blameworthy or praiseworthy. This is a strictly evaluative judgment. Ascribing responsibility to an agent or community of agents involves the problem of how to define in evaluative terms the action in question and the moral event it caused. An event or an action is defined morally with respect to a host of considerations, some of which we have considered thus far in this chapter.

We can isolate two aspects of evaluative judgments. First, one must make a judgment about the degree of culpability of the agent. An agent's culpability is lessened and assignments of responsibility are defeated if the agent acted under compulsion or out of ignorance due to no fault of his or her own. These excusing circumstances focus on a presupposition about rationality, rather than voluntariness, basic to responsibility. The presupposition can be clarified by saying that a responsible agent's actions must express intention and deliberation. Intention is "a kind of conceptual design present in the agent, the plan of what he aims to realize."[21] To be sure, we do not really have access to the designs present "in" other persons. Yet when we are seeking to ascribe responsibility, some determination of the agent's intention and whether or not what he or she designed to bring about was morally right or wrong is important. If we determine that the agent acted unintentionally, or that he or she did not fully understand an intended course of action, this will factor in the ascription of responsibility. A particular action might have more effects than the agent intended or could foresee. This was Aristotle's point noted before about ignorance making an act involuntary. What is morally important, therefore, is that the intended action and its direct effect be morally right.

The question of intention is linked to deliberateness. An agent deliberates about an action in terms of whether it should be done and also how it should be done. Valid ascriptions of responsibility require some judgment about the extent to which an agent (1) was able to deliberate rationally about the actions in question, that is, acted deliberately, and (2) chose the right means to the end in question. These are conditions of rationality necessary for genuinely responsible action. In Aristotle's terms, deliberation concerns the *manner* and the *means* of realizing some end.[22] This clarifies other excusing factors. If it can be determined that an agent was unable to act deliberately, say because he or she was rationally impaired either mentally or by being overcome by passion, provocation, or some intoxicant, then responsibility might be lessened. Similarly, an agent might choose the wrong means to an otherwise praiseworthy end. If it could be shown that the agent could be excused for this error, then responsibility would again be lessened or defeated. In each instance, the ascription of responsibility entails evaluative judgments. The judgments are evaluative in the respect that the one ascribing responsibility is determining whether or not an agent's intentional and deliberate action as well as the moral event the agent helped to bring about was right or wrong, good or bad. And this means, as I have noted before, that the ascription of responsibility, as well as the assumption of responsibility, rests upon some evaluative standard.

What makes the ascription of responsibility a distinctive activity is that a connection is made between causal and evaluative judgments about actions. If this connection is severed in the act of ascribing responsibility, then an agent could be (1) held responsible simply because he or she was the cause of something, or (2) judged even if he or she did not cause the moral event. The activity of ascribing responsibility is precisely the process of linking causal and evaluative judgments with respect to the role an agent or community of agents plays in bringing about a moral event. Given this, practices of ascribing responsibility help to constitute a moral community. They contribute to or detract from the good of human social life. This is why public opinion, systems of law and justice, and routine patterns of interaction between persons have such moral importance. What is at stake is our lives together.

Much of what we have just said about the *ascription* of responsibility is true as well of our acts of *assuming* it. Assuming responsibility for oneself also links causal and evaluative judgments. But our sense or internal perception of that link is not the social practice of praise and blame or systems of law and justice; it is the conscience. Conscience "testifies about" what I have done, or intend to do, with respect to some evaluative standard. The sting of conscience, as traditional Christian moralists called it, was the self convicting itself of its wrong with respect to actual or proposed courses of action. Through conscience, the individual internalizes the ascriptive practices of a community and its moral standards, even as the individual can and must evaluate the validity of those practices and standards. Conscience is to the self what the practice of ascribing responsibility is to the community; in each case there is a link between causal and evaluative judgments with respect to the identity of the agent. The overly sensitive conscience is one which holds the self responsible even when there are valid reasons to defeat assignments of responsibility.

The assumption of responsibility clearly entails causal and evaluative judgments. Rather than rehearsing that discussion with respect to the individual agent, it is important to isolate differences between assuming and ascribing responsibility. I have already mentioned one: it is possible, and often morally praiseworthy, to assume responsibility for actions we did not do. If my son breaks the neighbor's window with his baseball, I am responsible for the damage. More profoundly, Christians believe that the suffering of the world is borne by Christ, rightly imputed to him because as the Christ he has assumed it unto himself. This form of assuming responsibility falls outside of the demand of causal judgment applicable for the assignment of responsibility.

The validity of such representative action is dependent on two factors. First, whoever holds an agent responsible must be willing and able to accept the transference of responsibility. I can only pay for my son's damage to my neighbor's window if this is acceptable to my neighbor. That decision usually rests on a factual judgment about me and not the action: that I am capable of bearing the responsibility. My neighbor will not ask me to pay for the window if he knows I am broke but my son is not. This means, interestingly enough, that representative action requires power, not weakness as critics think. One must have the power to act in a way that is acceptable to whomever receives the transfer of responsibility. This is why in scapegoating mechanisms the "victim" is actually endowed with massive social and psychological power. Thus, my assumption of responsibility for another is contingent on my capacity to fulfill an action, my power, and also on conditions, like the acceptance of the transfer, beyond my control. This is so even if, as is often the case, I might *feel* myself responsible no matter what happens. Second, the assumption of responsibility for another must be voluntary, it must be an act of freedom. If it is not, it is scapegoating – the phenomenon in which the guilt of an individual or community is shifted to some "victim" who is then punished, sacrificed, or ostracized for an act or adverse outcome. Mechanisms of scapegoating not only impute power to the victim, they also violate the freedom of the one expected to assume responsibility and are deceptive with respect to causal questions which invalidate ascriptive judgments. Thus, the validity of the assumption of responsibility for others rests upon the freedom of the one bearing the responsibility and the acceptance of whoever or whatever is holding one responsible. An agent may rightly refuse to assume responsibility for others if either of these conditions is lacking.

The assumption of responsibility by an individual differs from the ascription of responsibility in terms of the demand for a strict causal judgment. Yet it also differs from ascription with respect to evaluative judgments. Whereas ascriptions of responsibility rely upon, fortify, and sometimes change social standards of moral evaluation, the assumption of responsibility does not merely rely upon those standards or seek to change them, but can be fundamentally at odds with them. In the name of assuming responsibility for self and community, an agent may claim that this requires resisting or objecting to shared moral values and norms. As Martin Luther King, Jr. and many others have argued, a person may be called to violate accepted social mores or existing laws in the name of some higher law or good. While it is the case that the protestor might be ascribed responsibility for his or her civil disobedience, she or he can bear that punishment and yet not assume guilt for that action. The "illegal" or supposedly "immoral" action might itself be seen as virtuous. For instance, Christian theologians like John Howard Yoder argue that Christians are called to represent to others a vision of human peacefulness in the midst of a violent world.[23] This stance puts the Christian community at odds with the political community. Christians reject operative social standards in the name of assuming responsibility for their lives under the demand of truthfulness to the Gospel. In each of these cases, the assumption of responsibility is based on a different evaluative standard than that used for the ascription of responsibility. The Christian or the protestor might – and probably would – submit to the punishment of the State, but with the proviso that this submission is not an admission of guilt. It too is a protest against the prevailing moral vision.

The assumption of responsibility can then diverge from patterns of ascription in terms of causal and evaluative judgments. However, determining the propriety of assuming responsibility turns on many of the same considerations as does the practice of ascribing responsibility to others. The assumption of responsibility also links causal and evaluative judgments. This "link" is not definable in terms of a social practice as was the case with the ascription of responsibility; rather, it is found in the identity of the moral agent. It is, as I have called it, conscience. The responsible individual is inclined "to a general manner of proceeding in all actions: the general manner characterized by consideration, conscientiousness and commitment."[24] This means that if I am to be a responsible person, then "I shall think about what I am doing, rather than act thoughtlessly or on impulse, and act for reasons that are faceable rather than ones I should be ashamed to avow."[25] A major dilemma facing any person is to grasp what it means to be responsible for oneself. How is it, in other words, that we assume our own moral identity? In spite of the critics of inwardness and conscience, critics explored in Chapter 1, can we still make sense of these ideas about persons basic to a Christian ethics of responsibility? I will address this question in the following chapter.

The assignment of responsibility in its two basic forms is concerned with agents and the moral events they bring about. Yet it is also the case that ascribing and assuming are themselves moral events, they contribute to, or detract from, the human good. They do so in special ways. The ascription of responsibility enacts the identity of some moral community and its beliefs about the human good; the assumption of responsibility by individuals is the enactment of personal moral identity. In this respect, the ascription and assumption of responsibility aid in bringing about or destroying a "higher" self or better form of community. In so far as this is the case, these are matters of freedom intrinsically linked to some vision of the human good and how we ought to live. This is why systems of justice and punishment are matters of responsibility.

It is this insight about freedom and the good which I want to elaborate in some detail in the next chapter. By doing so, I will offer an account of moral identity from the perspective of an integrated theory of responsibility. The present chapter has cleared the way for that discussion by exploring the connection between freedom and the assignment of responsibility. This connection, we now see, is important for any moral justification of punishment, or practices of praise and blame, and even a sense of ourselves as moral agents.

CHAPTER 7

Responsibility and moral identity

This chapter develops an account of moral identity from the perspective of an integrated ethics of responsibility. It builds on the previous chapter in that I examine how persons assume and ascribe responsibility for themselves. The formation of moral identity is through what I call the act of "radical interpretation." By this is meant the testing and transformation of the values and norms a person or community endorses as important to its life by some event, idea, or symbol which deepens the sense of responsibility, that is, the sense that the integrity of life ought to be respected and enhanced. This act of understanding is how the imperative of responsibility becomes constitutive of a person's or community's identity. The idea of radical interpretation links the theory of value, the imperative of responsibility, and a theory of agency in an integrated ethics of responsibility.

In traditional Christian ethics the link between person, moral norms, and a theory of value was the idea of conscience. Radical interpretation is a hermeneutical account of conscience. This account is crafted to meet the challenges of critics of Christian inwardness and responsibility, critics noted earlier in this book. In order to make this argument about moral identity the chapter begins by isolating levels of values found in persons and complex social systems. It thus expands on the theory of value outlined in Chapter 5. It then turns to current work in ethics on moral identity and responsibility. The chapter concludes with an agentic-relational account of human beings and also institutions developed through the idea of radical interpretation.

THE DOMAIN OF VALUE IN PERSONAL AND SOCIAL LIFE

The term "person" denotes the value of the individual with respect to the multidimensionality of goods which permeate life. Persons are internally complex creatures struggling for wholeness in relation to others. In so far as responsibility means respecting and enhancing the integrity of life, this requires that we respect and enhance persons. Similarly, societies are not simply aggregates of individuals but are composed of complex interactive subsystems with distinctive purposes, media, and values. Social responsibility involves respecting and enhancing the common good as the integration of social goods. In order to clarify these points, let us explore the domain of value in personal and social life.

At its simplest level, a person is a human being, and, more specifically, the living body of a human being. To violate a "person" is in large measure to violate bodily integrity and privacy. A person is someone who can be seen, touched, heard, and encountered, and, thereby, is experienced as present with us in the full texture of our existence in the world. We are relational beings. Of course, it may be the case that we are all in some respect a "doubting Thomas" with regard to the reality of others, and, oddly enough, even ourselves. Recall that the disciple Thomas would not believe reports of the bodily resurrection of Jesus unless he saw and touched the risen Christ for himself (cf. Jn. 20:24–29). So too, we apprehend the reality of the other person, know the other person as different than ourselves and yet with us in the world, through sight and touch and hearing, that is, by literally sensing the other's presence before us and with us. In this respect, bodily encounters are the medium for apprehending value. Similarly, a basic sense we have of ourselves as agents is in the simple exertion involved in physical movement. I have the odd sense that these fingers which I am now moving with such difficulty over the computer keyboard are mine, are me. But it is also the case, as with "doubting Thomas," that we sometimes must "get a hold of ourselves," touch our own bodies, to apprehend that these fingers are indeed me and not something other.

It is thanks to our bodies that we are situated in the world as relational, vulnerable beings. And the labor involved in bodily movement and encounters with others develops in us the sense of human efficacy. These facts ground the domain of pre-moral value, as I called them in Chapter 5. These goods, like bodily well-being and physical excellence, space and privacy, food and shelter, and aesthetic delight, radiate from our bodily existence in the world. These goods, as well as their correlate disvalues, are felt and experienced in terms of senses of pleasure and pain, attraction and aversion, which permeate the whole of our lives. Pre-moral goods are not subject *to* choice inasmuch as their value does not depend on a discrete act of will. The sense of these goods and disvalues testifies to their worth prior to choice. But pre-moral goods experienced in the bodiliness of existence are subjects *of* choice; we must respond to these goods. For example, one has to decide how, to what extent, and for how long to delight in pleasure or to endure pain, to surround oneself with what captivates or to flee that which revolts. In this way the feelings of pleasure/pain and attraction/aversion which permeate bodily existence are basic motives for action. Responsibility arises from embodied existence in the world with others. Our bodily existence mediates to us at the level of basic feelings a world of values to which we respond and in relation to which we exercise our capacities for action.

It is not surprising in this light that classical moralists like Aristotle and Thomas Aquinas defined "moral" virtues with respect to "a natural or quasi-natural inclination to do some particular action."[1] More specifically, moral virtues have to do with what Aquinas calls the appetitive power of the soul, that is, with what moves all human powers to act. Aquinas further argues that it is pleasure and pain, attraction and aversion, which are basic inclinations. For example, the virtue of courage is the excellence of character necessary to stand one's ground when one might naturally flee in fear from some person, possible event, or state of affairs. Temperance is the virtue necessary to moderate our appetites for what is pleasurable. My concern is not to rehearse the Thomistic theory of virtue. Rather, this comparison stresses the fact that our bodily existence, and thus our feelings of pleasure and pain, attraction and aversion, mediate a world of values and disvalues about which we must make choices about how to act. And this too confirms Aquinas' basic point. For he insists that "a habit of choosing, i.e., a habit which is the principle whereby we choose, is that habit alone which perfects the appetitive power."[2]

Thus, by a "person" we mean most basically an embodied human being who, in principle, can be responsible for her or his incarnate life. This definition is correlate to pre-moral values. To respect and enhance persons means, at its basic level, concern for the range of pre-moral goods manifest in physical existence. Human communities likewise depend on natural, finite conditions for their existence. This is true not only in terms of the interdependence of human existence with the natural environment, a theme we will explore in Chapter 8, but also in that all institutions rely on human labor, procreation, sustainable personal relations, patterns of fidelity, and communication. The concept of the common good, or social integrity, requires respecting and enhancing the natural conditions for social existence, including the lives of individuals and their basic needs.[3] As Roman Catholic moralists have argued, the individual necessarily has dignity and worth outside of social utility.

This brings us to the next meaning of person. We also speak of person in terms of social roles. As noted in Chapter 3, the term "person" is derived from the Latin *persona* meaning "mask." It is closely related to the term "character," another theatrical term. A person is one who plays a specific role; she or he is identifiable as a character in some complex series of actions and relations. The individual can and must be held responsible for actions in terms of the role(s) he or she ought to play. There must be some legal, institutional, or moral means for calling persons to account for their actions. We can examine social practices of praise and blame and the conditions which they stipulate as necessary for an agent to be blameworthy or praiseworthy. The question of personal responsibility is accordingly phrased in terms of one's office or role. I am not simply present in the world as a bodily being; I am a person defined by roles and relations. And this is, of course, the point of social theories of responsibility.

This sense of the person designates the ways in which an individual can be identified with respect to the roles which shape social relations. It implies, in other words, that certain obligations, duties, and forms of care for others are implicit in assigning a role to a person or in assuming it for oneself. To be sure, what those duties, obligations, and forms of care are and what they entail take different cultural expressions. To call someone "mother" in one culture does not necessarily designate the same set of expectations as it does in another culture. But the fact of descriptive relativism does not deny the basic point we are making. When we speak of a person in this second sense, we are designating the individual in and through social roles and the expectations these roles involve. A person's character is shaped and directed by the goods, duties, and expectations implied in the roles ascribed to or assumed by that person. A "good" mother, for instance, is the individual who in her character and conduct enacts the meaning of that role.

The term "person" designates a role-bearing individual whose identity has been shaped by assuming the values, obligations, and duties entailed in his or her role(s). The very idea of social roles entails a matrix of social relations and values. I can clarify the point through another comparison with classical thinkers. For Aquinas, the virtue uniquely concerned with relations to others is justice. Unlike other "moral" virtues, justice centers on what is due to others. As Aquinas notes, it is reason putting order into (human) operations, that is, into the working of the individual person or the working of a community.[4] And that order is simply that to each be rendered what is rightly due. The meaning of the self is the pattern of such relations within the community and its diverse roles and functions. Justice arises out of, and expresses, the deeply social and interpersonal nature of individual lives.

This idea of the self allows us to isolate other motivations for action. As Peter French notes, in forms of responsibility ethics which center on the identity of persons given through their social roles, "the primary motivation is to measure up – not to be found inadequate to the tasks that define one's identity."[5] Persons internalize some picture of goodness, say of the good friend, and then seek to enact this ideal in life. As I argued in Chapter 4 on social theories of responsibility, the drive to measure up to internalized ideal models involves the dynamics of building and holding self-esteem. We esteem ourselves according to the success of our actions as defined by our roles and relations, our practices and plans of life. Esteem is a distinct self-interpretation mediated by the evaluation of action with respect to social roles. I esteem myself as a friend, for instance, in so far as I enact my community's ideal of friendship. The debt of justice is that such action should be imputed or rendered to me. The inverse of esteem is shame. Shame means to "cover or hide oneself," which, if we recall the myth of Eden, is the experience of Adam and Eve when they fail in their obedience to God and their role as stewards of the garden. Shame relates to "failure, shortcomings, feelings of inadequacy and inferiority, exposure of weakness, and fault."[6] It is to be stripped of one's identity due to a failure to measure up to one's role and thus not to render what is due others. In terms of moral motivation, then, the desire for esteem and the dread of shame move our lives as role bearing individuals. The difficulty is that esteem and shame are bound to the ways in which a culture understands and values certain roles. Thus, distinct social goods, as I call them, are tied to this dimension of personal existence and the motives of esteem and shame which permeate life.

Here too we find social analogies. All human societies are functionally structured by complex patterns of rule governed systems.[7] These social subsystems range from economic activity through legal practices to communication or media systems. These functional systems are exceedingly complex internally and in terms of their interactions. Yet each system includes rules and mechanisms for giving and withdrawing recognition and regard to those in the system. In capitalism, for instance, recognition and regard for agents is based on the flow and acquisition of goods and capital. The esteem or shame of economic agents is keyed to this mechanism of giving or withdrawing regard and recognition. The poor, we know, often internalize the shame of their poverty. Similar mechanisms function in the legal, media and other social systems. In religious terms, these systems have a cultic function: they give and withdraw recognition and regard and thus are the context of esteem or shame in the community. Accordingly, the primary virtue of the social system, like the social good of the person, must be justice. There is a demand in each of the subsystems for fairness based on recognition and regard for the claim to esteem, or shame, by persons in the system. Thus, the common good is not only a matter of respecting and enhancing the material conditions for life, it also entails the demand for social and civil justice.

This brings us to a third domain of personal value. What we mean by "person" is not simply bodily existence in the world or the social roles persons adopt, but also someone who can be linguistically addressed and designated. One designates persons in different ways. We make a distinction between a speaker or actor, the person addressed, and also the individual or thing spoken about or the act done. There is an oddity in this, however. For instance, the sentence "I love my son" can be uttered by any parent of a male child. The sentence purports to designate a specific person (W. S.) and a particular relation of love (my son, Paul). And yet the sentence in terms of its form and structure effaces the particular person by its applicability to anyone who utters it. In uttering a sentence which can have widespread, nearly universal, application, the individual person, oddly enough, means to designate the distinctiveness of her or his life. To recall another example, when I say "it was I who broke the promise," I designate myself as a person through a grammatical form (the first person pronoun) applicable to anyone who can use this language. In uttering the sentence, who I am appears, as it were, as the link between a linguistic designator ("I") and a particular deed (breaking a promise). This is important for our capacity for reflexive self-understanding and self-designation. Yet under the pressure of linguistic analysis it also threatens to reduce the self, the "I," simply to a function of language. Who I am is dependent upon the language I speak or which bespeaks me.

The capacity for self-designation is basic to what we mean by a person. As I argued in Chapter 1, without this capacity one could never designate oneself as an agent and thus a member of the moral community. Yet the ability to designate oneself as an agent with respect to the linguistic resources of some moral community depends on the other meanings of person noted. The power to designate oneself relies on the person as a bodily being acting in the world within some community and its patterns of responsibility assignment. Yet this linguistically constituted sense of person does denote the level of reflective goods we noted in Chapter 5. The capacity to reflect upon my existence is correlate to designating myself, or being so identified, as a responsible agent. At issue then is the truthfulness of our lives, that is, the extent to which our actions and relations are true to the ground projects which constitute our moral identity. It is at this level of personal existence that matters of guilt, self-respect, and also love of self and others find expression. We only have these feelings in so far as we are persons in this reflective sense. The fact of these feelings, as I noted before, is basic to attitudes we regularly hold about agents. This is a form of conscience. I know myself along with knowing (con-scientia) the rules, beliefs, values, and moral discourse of a community through which I designate myself as a self. Conscience in this sense – let us call it conscientiousness – is the internalization of the values and ascriptive practices of a community and is thus a judge of personal conduct.

We are then concerned with self-description and the institution of language as a domain of value in social life. By language I mean two things. First, language refers to the linguistic code as a system of signs through which meaning is produced. As post-structuralist thinkers insist, language is "a vast system or sign structure whereby meaning is determined by a mutual relation of signs which transcends the localized talk of individual speakers."[8] In a profound sense persons belong to language in so far as their acts of self-designation rely on the linguistic resources of some community. This is a central point of social theories of responsibility. But language, second, also means discourse: the act of somebody saying something to someone about something.[9] There is a human being, a person, involved in all communication. The capacity for self-designation, and thus reflective goods, depends on the linguistic code but also breaks open the code to actual personal life. The linguistic practices of a community must respect and enhance the conditions necessary for responsible existence and that involves the means for reflective values.

Language is one of the greatest social goods; it is also the medium for the distortion of truth and life.

This brings us to the last sense of the term "person" important for an integrated ethics of responsibility. By person we also mean the personality of the human being, the self. The person is bodily present in the world, adopts and assumes various social roles, and can designate himself or herself and be designated by others through linguistic practices. Many theorists have argued that it is possible simply to define the self in terms of one of these meanings of "person." But the self is not reducible to these, or to their coordination in any specific action. The person can, and must, adopt some posture, make some evaluation, of his or her bodily life, social existence, and patterns of identity-designation. As Susan Wolf has put it, we have the "ability to step back from ourselves and decide whether we are the selves we want to be."[10] As I argued in Chapter 3, the capacity for self-criticism is at the root of moral freedom and also how we become responsible for ourselves. A person's ability to be self-critical is what gives us the sense of depth, texture, dignity, and also burden to our lives as bodily, social creatures. This is what Christian ethics properly means by conscience. Conscience is the living reality of the self in its full moral complexity apprehending the integrity or fragmentation of its life and its community before the good. This is why Christian theologians have claimed that since God is the good, conscience is a sense of the divine.

Interestingly, the language of responsibility from this perspective designates something about the temporal character of the self. We understand ourselves, we have our moral identities, as historical agents in relation to others and the world. Persons exist as selves in a moral space of relations through time. Thus, it is ordinarily believed that a person is and ought to be responsible for the consequences of his or her actions. Assuming responsibility, as noted in Chapter 6, has a retrospective and future oriented character to it. And it is also the case that a person is responsible in the present. Being responsible entails a commitment to self-constancy through time with respect to actions, intentions, and consequences of actions. As Roman Ingarden has noted,

In order for the agent to be able to bear responsibility and also to rid himself of it, his identity, especially the identity of his personal being, must persist, despite all transformations which may meanwhile occur in him, and it must be maintained for as long as it takes him to get rid of the responsibility.[11]

The persistence of personal identity through time is a matter of dispositions, habits of belief and thought, customary actions, and also commitments. For what is at issue is the coherence through time of the diverse goods rooted in the levels of personhood. Without the persistence of identity through time, without some conception of the "self" bound to memory, hope, and capacities for self-reflection, the idea of responsibility becomes vacuous. Thus, at this level we are concerned with the unique good of moral integrity.

And what of the domain of social value? Communities exist through time as cultural traditions. Tradition, from the Latin *traditio*, is the handing on of beliefs, values, and customs. A tradition has a unique temporal structure, as Hans-Georg Gadamer has argued.[12] In any moment, the past shapes contemporary consciousness and opens possibilities for the future. Personal life draws its substance from the mores of a tradition or community. This does not mean that these values impose themselves heteronomously upon us. We can assess the values which have shaped who we are and thus orient our lives and our actions. But in doing so we put ourselves into question in the most radical sense of the term. The paradox is that, in this activity of questioning, a moral tradition is deepened and extended by enabling its members to respond to new questions and problems. A tradition, as Alasdair MacIntyre has rightly insisted, is not a closed *depositum* of beliefs; it is more like an open-ended debate about the human good.[13] The persistence of a tradition rests on its members sustaining an unending debate about the good. The responsibility of any community, like the Church, is to sustain debate about its moral purposes, so that one can designate its existence as a moral tradition. That is to say, one can speak of the moral integrity of a community only to the extent that the community sustains and is sustained by reflection on moral purposes.

When we speak of the person we mean an individual, bodily self who lives amid complex social relations and roles which contribute to her or his identity and who can also designate herself or himself, and be designated by others, as the same person through time. The idea of a person denotes the integration of bodily existence, social roles, and constancy through time in communication with others. Each person manifests the variety of pre-moral, social, and reflective goods we ought to respect and enhance. This means that we ought to respect and enhance persons. The same would seem to be true of communities. Any community manifests within itself the domain of values which are to be respected and enhanced. The responsible person or community is unique in that life is integrated with respect to the demands of responsibility. How are we to understand this? How do we assume responsibility for the multidimensionality of our existence as persons? Can we speak of corporate responsibility? In order to address these questions, I turn to work in ethics on the nature of moral identity and responsibility for self.

EVALUATION AND RESPONSIBILITY FOR OURSELVES

I have been arguing that responsibility is tied to our sense of self and to the life of communities. I have further argued that the capacity to assess the projects which give meaning and purpose to one's life and the life of a community is especially important since it aims to integrate in a distinctively ethical way values found in personal and social life. How are we to think about this unique form of responsibility? We can begin with a common experience.

All of us, I trust, acknowledge that it is important to identify with the values, goals, and desires we hold, as well as with the actions we undertake. This identification cannot simply be forced on us. We have the capacity to decide whether we are the selves we want to be. The capacity for moral self-transcendence, while never completely free of self-interest, is basic to the constitution of moral identity. A person's identification with, and judgments about, desires, values, goals, and actions is the root of her or his self-determination. We feel violated when that identification is denied, since without it one is deemed mad or incompetent, not responsible. Similarly, our sense of responsibility is violated when someone or something other than ourselves attempts to force us to accept values, goals, and desires not our own. The fact is, then, that freedom is neither a matter of brute choice nor simply determined by inclinations. We can decide which desires, affections, and loves we want to shape our character. This is basic to moral responsibility. Charles Taylor has recently explored these matters and proposed a version of what I called in Chapter 6 an evaluative theory of freedom. I will draw on Taylor's work in developing the idea of radical interpretation within an integrated theory of responsibility.

Taylor argues that what constitutes us as moral beings is that we can undertake the "radical evaluation" of our lives.[14] We are not formed as selves in discrete acts of choice or brute acts of will, as existentialists like Sartre want us to believe. All of our choices depend on the values we endorse and the various social and historical sources which have funded the modern sense of self. We cannot choose not to have some values which direct our choices. We would not have a commitment to morality at all if we were not formed in certain ways. As Taylor puts it, we orient ourselves in the moral space of life by some idea of the good, some strong evaluation. This is not a simple psychological observation about persons; it is an ontological claim about human existence. We are those beings whose existence is a matter of concern to us and thus we orient our lives by some framework of value.

If we are properly to understand ourselves we must attempt to grasp the various, and often conflicting, values that have shaped our identities. To live without concern about which values ought to guide our choices is to live at best an impulsive life and at worst a chaotic one. The moral life requires that we consider what we fundamentally value. Because traditions and communities have shaped what we care about, to question those value orientations, to engage in radical evaluation, is to submit our lives to self-criticism. This kind of self-criticism allows us to formulate what Harry Frankfurt has called "second-order" desires and volitions. As opposed to the "first-order" desires and volitions we simply happen to have, "second-order" desires and volitions are those we want to define who we are. All of us have first-order sexual desires, for instance. The question is whether or not we formulate second-order desires and volitions which entail some judgment about those basic desires. And this capacity is basic to what we mean by a self.[15] Second-order desires and volitions are not given in the emotional fabric of our lives. They are the ones we consciously want to guide and orient our lives as persons. In so far as we come to endorse in our decisions the values that we want to orient our lives, we are responsible for ourselves, we are autonomous moral agents. This argument is a variant of the claim I made in Chapter 5 about lower- and higher-level goods. A higher-level good, like moral integrity, is for the agent a second-order desire and volition with respect to first-order desires and values.

Taylor's idea of radical evaluation and the point about second-order desires and volitions are important for an integrated ethics of responsibility. They help us to see that we come to moral situations with the values of traditions in hand and also with first-order desires and volitions. But these ideas also provide the conceptual means to explore how we are responsible for ourselves through self-criticism. According to Taylor's theory, to be autonomous is not to act out of radical choice; it is critically to endorse the values we want to orient our life. We are indeed free because we make critical evaluations and form second-order volitions and desires about who we want to be. But this freedom is compatible with the social and historical character of human life and the linguistic means we use to depict our lives. Responsibility is linked to our capacity to reflect on and then revise or transform our lives through criticism of what we care about. Radical evaluation is what Taylor means by this process of self-revision.

What standard should we use to evaluate our values? When we ask about the standard for evaluation, Taylor's project becomes confusing. He writes:

Our attempts to formulate what we hold important must, like descriptions, strive to be faithful to something. But what they strive to be faithful to is not an independent object with a fixed degree of evidence, but rather a largely inarticulate sense of what is of decisive importance.[16]

Taylor is right to insist that moral descriptions strive to be faithful to something; as I argued in Chapter 5, they have a realistic intention. Therein lies the problem in his argument. How can our *moral* evaluations be faithful to what Taylor admits is our "deepest unstructured sense of what is important"?[17] Taylor's argument at this level threatens to devolve into subjectivism in the name of the contemporary value of authenticity or truthfulness to self. He tries to answer this problem by arguing that radical evaluation must deepen our recognition of the good. Only a recognition of good can ground principles about right conduct, like justice, and avoid subjectivism. But this poses a question. "Do we have ways of seeing-good," Taylor asks, "which are still credible to us, which are powerful enough to sustain these standards [of justice and benevolence]?"[18] What is at issue is not the specification of a good that all persons must seek, a specification not possible in pluralistic cultures. Rather, Taylor is asking about the moral perception and motivation basic to any way of life.

In Taylor's argument, the seeing-good basic to any commitment to justice and benevolence which avoids subjectivism is inseparable from Christian *agape*, a love that endorses the claim in Genesis 1 that "God saw that it was good." This love acknowledges that others are worthy of care and respect independent of the self's wants. Seeing the good is more basic than choice even in the commitment to justice and benevolence as paradigmatic moral values. But Taylor is not arguing for classical realism. We cannot circumvent the human knower and valuer in a theory of moral knowledge and value. Our perception of the good is not devoid of our own creative act of making. In Taylor's words, "What we have in this new issue of affirming the goodness of things is the development of a human analogue to God's seeing things as good: seeing which also helps effect what it sees."[19] The *agapistic* perception of goodness cannot circumvent the self who perceives and acts. "Put in yet other terms," he writes, "the world's being good may now be seen as not entirely independent of our seeing it and showing it as good, at least as far as the world of humans is concerned."[20] Only this kind of seeing and showing good in the struggle to enact goodness in life can sustain us in our commitment to justice and benevolence. This is the point, recall, of hermeneutical realism.

Taylor explores responsibility for self with respect to the radical evaluation of desires and wants. The idea of personal responsibility hinges on an act of self-criticism in which an individual assesses his or her character and commitments with respect to something taken by him or her to be what life is about. But from the perspective of an ethics of responsibility, moral criticism must be defined with reference to what demands and empowers some answer from us in the claim of its worth actually to exist in the world. This means that moral identity is constituted not only by what we care about and its evaluation, as Taylor insists, but also by the experience of respect for others. Respect is the recognition of and regard for what is other than the self and its projects. The experience of respect discloses the fact that we are always interacting with others and our world. And this is because "respect rests on the recognition of something as intrinsically worthy of it."[21] We have experiences of value because we are beings who care, but such experiences can take the form of moral respect only because of a primal recognition of others in the world. Care and recognition are multiple sources of human action.

I can clarify that care and recognition are basic to moral identity through a distinction between movement, action, and moral encounters. For a movement to be an action it must be directed to some end apprehended as good, otherwise we would be speaking merely of motion. This was the point of Aquinas' and Aristotle's arguments about what is "voluntary." A voluntary act is one moved by an intrinsic principle of movement to an end. If we are to be responsible for ourselves, we must then critically evaluate the end or good that orients our lives as valuing, caring beings. This is the force of Taylor's theory of radical evaluation, and, as far as I can see, every evaluative theory of freedom. Yet while examining the conditions of distinctively human action is important for genuine self-understanding, it is not sufficient for an account of moral identity. This is because recognition of the other is also a primal source of moral motivation. The value of others is ꞏncountered in such arresting moments as being horrified over gang violence, being grasped by the claim of life in the face of a child, or beholding the beauty of a Wisconsin lake. These experiences of moral recognition help to explain the profound resentment, compassion, and also gratitude which characterize the moral life. Moral conduct presupposes conditions for action, such as the fact that we can evaluate what matters to us, but then insists that we recognize what is an end-in-itself despite what we might care about and value. An agentic-relational theory of persons must include both the analysis of what we care about and also recognition of the other as basic to moral conduct. There are multiple sources of human action, a claim I made in Chapter 6 about evaluative theories of freedom. But this means that care and recognition in this account are not grafted onto each other. One does not begin with what we care about and then graft on to this a theory of recognition of others. Care and recognition are equally basic experiences, because human beings, as relational, self-interpreting agents, are always acting for ends they value and interacting with others. Human inwardness and the drive to wholeness are born from the interaction of these sources of action.

A condition for responsible human action is, then, some critical evaluation of the good or goods that ought to orient our lives. Human action is teleological. Yet, we can isolate another kind of end which is basic to moral action. Here the end for which an action is done is not a state of affairs to be realized, or, as Taylor argues, a good that orients action in terms of what we value. It is something that is recognized to exist as a good-in-itself with a claim to be respected in all action. The concept of an "end" in this case is bound to an actual, existing being or state of affairs rather than something to be achieved, sought, or realized. It is an "end" in so far as the actual reality, the existent being or state of affairs, is acknowledged and respected not simply as a means to some other purpose but in all action.

The dilemma of our moral lives is that ends of these two kinds can and do conflict. What we care about can and does conflict with recognition of others. When it does, we engage, unless we are "wantons," in reflection about what we ought to care about and respect. In so far as one experiences this form of moral conflict, self-criticism would seem to be basic to moral identity. And rather than being, as Taylor thinks, an inarticulate sense of what is important, the principle of self-criticism would have to relate and order maxims of choice rooted in the deep experience of care and recognition of the other. As we have seen, this is precisely what the imperative of responsibility does; the demand that we respect and enhance the integrity of life formulates and orders maxims (respect, enhancement) to guide choices about actions and relations rooted in our sense of the two ends of human conduct.

On this account, the aim of moral self-criticism in Christian ethics is to make the source of human action, that is, what we care about, to be identical with the claim to respect the integrity of life perceived in others and ourselves. The responsible person or community is one who cares about what ought to be respected. This is actually what Christian faith means by love; *agape* is the bestowal, the gift, of care based on the recognition of the goodness of what is. Any apprehension of the goodness of existence which binds the self to the cause of life is implicitly an expression of this form of love. Thus, if we are to speak of moral identity in Christian ethics we need to explore how what we care about as the basic condition for action, is transformed in light of what we ought to respect. How ought we best to describe this transformation – even sanctification – of moral identity? I turn now to the idea of radical interpretation, a hermeneutic of conscience, as basic to an ethics of responsibility.

RESPONSIBILITY AND RADICAL INTERPRETATION

Let me review the steps of the argument taken thus far in this chapter. I have isolated the domain of value in personal and social life and how these values press toward the question of the moral integrity of the self or some community and its tradition. I then examined the work of Charles Taylor and others with respect to the question of how we are responsible for self. This requires the evaluation of the values which have fashioned our sense of self and thus our root commitments. The ability to assume responsibility requires the capacity to form second-order desires and volitions. What defines the moral agent is an activity of critical assessment of the ground project(s) to which the meaning and value of life is bound. There is no reason in principle that communities cannot have mechanisms for this kind of analysis, say in voluntary associations, public moral debate, and democratic elections. In so far as this is the case, we can speak of the moral identity of a community.[22] I return to this point below.

We have thus entered anew the question explored in Chapter 6 about moral freedom and its connection to conceptions of the good. I agree with Taylor and Frankfurt that we need an evaluative model of freedom. Yet I have also tried to show that the principle of self-criticism must be an imperative which formulates and orders maxims for choice rooted in the primal moral motives of recognition and care in terms of the integrity of values in personal and social life. These primal motives, I should note, need to be elaborated in terms of pleasure/pain, esteem/shame, innocence/guilt, and self-constancy/incoherence isolated above in the layers of values in personal and social life. That elaboration is not possible in this book. The point here is that since the imperative of responsibility is used to interpret life, it aims at refining and deepening moral experience, the sense of responsibility. For the Christian this means that the criticism of self and society through the imperative of responsibility arises from and aims at the discovery and deeper perception of the reality of God in the world. In order to clarify this, I want now to introduce the idea of radical interpretation.

Radical interpretation is reflective, critical inquiry aimed at the question of what has constituted our lives in terms of what we care about and what ought to guide our lives under the demand of respect for others. It is the form conscience takes in the lives of social, linguistic, self-interpreting agents. Such inquiry becomes "radical" when it strikes at the root of who we are, our identity-conferring commitments, and the conceptual frameworks that we have used to understand ourselves and our world. Radical interpretation is a way to articulate how moral identity is constituted and transformed through an act of understanding. We do not merely evaluate the moral worth of others in terms of our interests or inchoate feelings; we also understand the moral life and what we care about in terms of the experience of the recognition of others. Radical interpretation is an activity of self-criticism in which a condition of our very acting, what we care about, is transformed through the recognition of goods that ground the moral life and ought to be respected and enhanced in all our actions and relations. It is the practice of self-examination which fosters a sense of the moral depth and inwardness of life.

There are deep theological reasons for adopting the idea of radical interpretation as central to a theory of moral identity. In the biblical texts, the Hebrew prophets called Israel to engage in this kind of interpretation through repentance and remembrance of the covenant and God's fidelity. The people become Israel through remembering and living out the covenant as definitive of their identity rather than following false gods. Their sense of corporate responsibility is rooted not in a collective self, but, rather, in mechanisms of social interpretation of identity-conferring commitments. Likewise, the teachings of Jesus confront the reader with the same demand about the reign of God. When Jesus asks, "who do you say I am?" he forces his listener to undertake the act of interpreting what she or he cares about and respects. The hearer radically interprets himself or herself in response to the demand for some answer to Jesus' question. If Jesus is the Christ, then one ought to follow him as one's basic moral project.[23] The idea of radical interpretation also enables us to understand the divine name, God's identity, in the biblical texts. God names God's self and in doing so constitutes the identity of the community through commitment to justice and mercy.

The act of radical interpretation is the primal deed of genuine moral identity. It constitutes the self-understanding of an agent, or community of agents, who exercises power in terms of the experience of a claim of others and the world upon the agent. Radical interpretation is an expression of moral freedom since it informs and directs choices based on the critical assessment of value orientations with regard to what solicits respect. This enables us to affirm the degree of autonomy needed for the moral life. We can and must critically assess our value orientations and endorse only those that meet the test of interpretation. The idea of radical interpretation makes moral transformation, or self-revision as Taylor presents it, into an epistemological principle in ethics. Moral knowledge is a process of transforming care for and recognition of others into a settled disposition to live in responsibility; it is to see others as ends-in-themselves and this entails the demand to realize life in others and ourselves. The moral life is nothing less than the ongoing transformation of self through the refinement of moral experience and sensibility. This is what classical moralists would call the education of conscience.

However, if we are to specify the demands of moral responsibility, we must assess our actions and values by some idea or symbol which renders the recognition of the value of existence into an imperative for action. This requires not only refining moral experience in the face of its fragility and distortion but also defining a principle of choice. Radical interpretation enables us to articulate *how* the critical process of self-assessment takes place and thus constitutes moral identity. It does not provide an imperative for action consistent with this experience and self-understanding. To use traditional language, conscience apprehends the moral law, but it does not promulgate it. More specifically, conscience apprehends the imperative of responsibility that in all actions and relations we are to respect and enhance the integrity of life before God. The imperative, the moral law, is rooted in the nature of agents, we saw in Chapter 5, and ultimately in God as the source of responsibility (Chapter 8). Radical interpretation is the appropriation of this imperative into the self-understanding, the identity, of an agent or community. What has now become clear is that there are reasons endemic to human existence why this imperative can and must and may move persons to act. The deontological demand of respect and the consequentialist concern for enhancing the goods of life find their motivation in the primal experiences of recognition and care.

What is at issue theologically is the relation among these demands, their motivational roots, and understanding life before God. In the Christian tradition, ultimate power as it evokes gratitude and reverence is identified as "God." God is value creating power. God is the creator of the moral order of reality in so far as the mystery that is ultimate power names itself with respect to the goodness of creation, the demands of covenant fidelity and justice, and the redemptive power of love. Outside of this name and the history of interpretation it entails, the meaning of ultimate reality is the abyss of power. But Christian ethics does not take the fact of power as the central datum for interpreting the world. For Christian faith "God" is the name for the radical interpretation of ultimate reality in which power is transformed in recognition of and care for finite existence. "Who" God is, the divine identity, is interpreted with respect to specific values and norms: God is creator, sustainer, and redeemer. The reality of God requires then a specific construal of the world. More boldly, the identity of the God of Christian faith makes responsibility basic to our understanding of reality. We are to see and value the world and all that is in it in terms of the transformation of power to respect and enhance the integrity of life. This is how Christian ethics must respond to the current criticisms of mythico-agential accounts of reality noted in Chapter 1. And given this claim about God and world, the requirement placed on us as moral beings is first and foremost to endorse as constitutive of our own identity the subjection of power to norms of well-being. The act of radical interpretation undertaken in terms of the identity of God is the personal and social practice of theocentric conscience; it is how Christians seek to imitate the divine goodness in our lives.

The importance of this theological claim about the divine identity and what it means for self-understanding and a construal of the world should not be quickly dismissed. As we have seen in other chapters, the pervasive assumption of Western moral traditions is that the human good is not to be found in the release of power, but in the exercise of human capacities with respect to moral norms. This belief, expressed in the obsession with responsibility, grounds our sense of human dignity and also the insistence on justice and mercy in social life. These commitments signal the transformative power of theistic faith on moral consciousness, whether or not the religious commitment it entails is rendered explicit.[24] If we were to excise completely from Western consciousness this notion of the divine, or some functional equivalent to it, and its ongoing transformative power on moral sensibility, we would not simply be freed from excessive guilt and the weight of inwardness, as Shklar, Smiley, and others argue. More profoundly, there is reason to believe that the sheer release of power would indeed come to define the human good. As Friedrich Nietzsche saw in advocating this agenda, it is difficult to imagine what our world would be like in light of such a transvaluation of values.[25] In fact, no viable world worthy of our commitment would be possible. The creation of a viable culture, like becoming a self, entails the binding of power to a recognition of and care for finite existence and its future. This is the truth of the Christian theological perspective on the nature of reality expressed in the symbol of divine creation. I will return to this point in the next chapter of this book.

Explicit faith is a second-order desire and volition to understand reality and ourselves with respect to a trust in and loyalty to God. Interpreting life theologically enables a truthful insight into what is to be respected and realized and thereby transforms moral self-understanding. This insight can be formulated in two ways which bear on the actual conduct of life; these are traditionally precepts of conscience. First, it can be formulated in terms of the first precept of practical reason, seek good and avoid evil. This precept concerns a condition for distinctively human action, that is, what we care about. It means that responsible action seeks to enhance and realize finite life and to avoid its destruction. Second, the root moral insight can be formulated in the precept that in all our actions and relations we ought to respect, even reverence, life in relation to God. The principles for moral character and conduct center on a recognition of the claims of others upon us and an active concern for others and the world. These are precepts of responsibility consistent with the relation of recognition and care in moral experience when respect, as recognition of the other, and care, as a condition for action, are radically interpreted from a theological point of view. The imperative of responsibility must then include and order the above precepts: in all actions and relations we are to respect and enhance the integrity of life before God.

Radical interpretation provides the means to articulate how

moral norms become constitutive of moral identity and empower a way of life. It does so in so far as this act of reflection provides the means for agents, or communities of agents, to assess and to transform their self-understanding with regard to what they care about and the experience of respect. In terms of current work on moral subjectivity, radical interpretation reclaims the Christian idea of conscience as basic to a theory of responsibility and moral agency. And it means that the experience of goodness is achieved by interpreting ourselves and our world through Christian discourse about God. Radical interpretation does not mutilate human goods; it properly integrates them in terms of primal experiences of recognition of others and in terms of what we care about as decisively important.

MORAL IDENTITY AND CORPORATE RESPONSIBILITY

Before concluding our inquiry into the formation of moral identity we must address a final question. Does the idea of radical interpretation help us address the problem of the moral identity of human collectives?[26] Is it possible within an integrated ethics of responsibility to speak of corporate responsibility? Ancient societies, as noted before, held that a community as a whole could be held responsible. The idea of the scapegoat in the biblical texts relies on the belief that the sins of a whole community could be assigned to one person or animal for the propitiation of the sin (cf. Rom. 3:21–26; Lev. 16:11–22). This conception of corporate responsibility has eroded in modern society. The reason for this is not difficult to grasp: it is difficult to speak of a whole society as an intentional agent.

We can clarify this problem about responsibility and corporate agency by focusing on economic institutions because a good deal of the debate in ethics has centered on them. Can we really ask: "who" is the corporation? The most obvious retort to this question was put by Milton Friedman and recently again by Paul Weaver.[27] They argue that claims about corporate responsibility are confused because the corporation, after all, is an artificial reality, a legal creation which exists to make a profit. The assumption is that by contrast a moral agent is defined by having a self-identical center of consciousness, a "soul." But this "soul" is understood in a specific way within modern economics. As Amartya Sen has pointed out, the view of the human in much economic theory "is that every agent is actuated only by self-interest."[28] Such an agent freely determines her or his behavior while entering into constitutive relations with others through contractual agreements. These agents exercise their subjective preferences by seeking to maximize interests constrained only by those contracts with others. This idea of agency, coupled with a voluntarist conception of freedom, is a hallmark of much modern thought.

Once this portrait of agency was accepted by political economists and others, any attempt to reflect in other ways on economic life seemed foiled at the outset. It is difficult to speak of corporations as having a self-realizing soul. The critics of corporate responsibility argue that, unlike the consciousness persons have of being agents who enter contractual agreements subject to moral criteria (like a principle of justice), we cannot speak about corporations in these terms. For this reason institutions fall outside the scope of the moral community. The *locus classicus* for this argument in Christian ethics was Reinhold Niebuhr's *Moral Man and Immoral Society*.[29] His point was that only persons have the degree of rational and volitional self-transcendence necessary to qualify as true moral agents. In so far as morality depends on the capacity to transcend brute self-interest, societies are by nature immoral.

In response to these problems, various attempts have been made to speak of "corporate moral agency."[30] To do so requires showing that a corporation (1) is the cause of events that can be evaluated regarding intentions and outcomes (2) is free and purposive, or rational, about its activities, and, finally (3) is admissible to membership in some moral community to which it is accountable and whose beliefs and values are norms for judging actions, intentions, and outcomes. Thinkers have tried to meet the demands for specifying the agency of a corporation in one of two ways. First, ethicists like Peter French explore the decision-making structure in a corporation as analogous to human agency and thus center on the problem of consciousness. If it can be shown that this analogy holds, then we can speak of corporate agency. Second, other thinkers contest the "corporatist" argument and claim that only human persons can be *moral* agents, with all the rights and burdens this entails. To admit corporations to the moral community seems to demean human beings.[31] Granting the veracity of the critique of corporate responsibility, these thinkers concentrate on the decisions, virtues, and character of managers, the agents of shareholders. The manager has representative responsibility, as I have called it, within the corporate world. Regardless of the status of the corporation *per se*, at least managers are *moral* agents and thus can be held to canons of social responsibility relative to their shareholders, principal, and the public.

Must we grant the initial assumptions about agency found in the debate? Do we face a genuine either/or: either corporations are moral agents and thus human dignity is qualified, or only persons are moral agents such that corporations escape moral responsibility? In order to undertake a different way of thinking about agency, I have detailed an activity, the act of radical interpretation, within which moral *identity* arises and is rendered intelligible. The argument about agency hinges not on consciousness, or conscience, as something radically distinct from the activity of critically interpreting courses of action. An understandable identity, even if a distorted one, arises within acts of assuming and ascribing responsibility. Identity is bound to practices of responsibility assignment. The activity of radical interpretation as a practice of responsibility assignment constitutes the moral identity of an agent. To be a responsible agent is to examine one's life in this way as an exercise of genuine freedom. Through the complex act of interpretation, the identity of an individual is presented in ways that call for the evaluation of the agent's actions and relations to others. We need not assume anything about the actor's "soul" prior to this act of interpretation in order to assign responsibility. This was an insight of social theories of responsibility. What we must grasp, then, is that there is a process analogous to personal moral identity formation operative in communities in terms of their traditions. A tradition is the self-interpretation of a community or culture; a moral tradition, if it is viable, is the radical interpretation of a community. We might speak of this as the social conscience of the community. In the Church, for instance, it is found in the legacy of ecclesial reflection on the moral life. We have an answer then to the problem of how to speak of the moral identity of an institution or community. In so far as an institution represents its identity through interpretive practices (reports to shareholders, accounting practices, political traditions, social histories, communal memory) and thus constitutes itself through time, then it is rightly held responsible for respecting and enhancing the integrity of life.

This is an analogy between persons and communities. Human beings alone can and must render their own identity amid complexity with the risk of fault and estrangement as well as the possibility of forgiveness which this entails. Persons do so within the linguistic and temporal interpretive activities that engender identity. Only persons can *sense* the tension between their acting and their lives, a sense of conscience, that leads to confession, thanksgiving, or evasive denial. And only persons can *reflect* upon their sense of responsibility and their behavior while also being their own moral accountants. Because of this, our moral being is always in balance. The sense of this balance is what is meant by the "voice of conscience" as part of what it means to have a "soul."

Persons are capable of degrees of self-awareness, deception, and fault not open to communities and institutions, even if the destructive capabilities of social totalities far exceed that of individuals. Thus while institutions and communities in virtue of enacting some tradition of reflection are "agents" in the moral community along with persons, they are not equal members. Institutions do not enjoy the same rights and burdens as human beings. To demand equal respect would be to violate crucial moral differences. It would be to fail to understand the *analogy* at work in exploring personal and social identity. Persons must bear responsibility for themselves in a way characterized by reflection and affective awareness. Institutions enter the moral community only in so far as persons continually question and assess the purpose of the institution and thus hand on a tradition of moral reflection.

Personal and social identity is formed and assessed through acts of radical interpretation founded on a commitment to some moral project, some orienting faith. This means that self and community are always measured by a good which transcends immediate existence, the good of the integrity of life. If the argument of this chapter is right, all communities and individuals are bound by this imperative and thus required to interpret their existence through it. This follows since, as we have seen, persons and social systems manifest in their own existence the complexity of pre-moral, social, and reflective values. The denial of personal and corporate responsibility is nothing less than the denial of the conditions necessary for coherent life. And that choice is not one which ethics can either understand or justify.

CONCLUSION

This chapter completes Part II of the book. By developing the idea of radical interpretation as a hermeneutical account of conscience with respect to the domain of value in personal and social existence and the imperative of responsibility, the theory of morality and the theory of the nature of moral agents have been linked in an integrated ethics. In doing so, I have presented an evaluative theory of freedom within a hermeneutical realist account of value while recasting Christian discourse about conscience. Conscience is not a faculty of the soul, a divine spark in the mind; it is the practice of radical interpretation within which personal and social identity is constituted and formed in terms of the imperative of responsibility.

In this chapter I have also begun to introduce the importance of a theological point of view in ethics. The next part of the book turns directly to theological questions with respect to the threat human power poses to the integrity of life on this planet. Thus, I conclude the book with a radical interpretation of our current social situation and thereby return to themes introduced at the outset of this inquiry.

III

The source of responsibility

CHAPTER 8

Power, responsibility, and the divine

A defining feature of current life is the radical increase in human power over the last few centuries. Human power now can alter the genetic structures of life and it threatens the environment. In this chapter, I explore the source of responsibility in order to counter the belief that power is the ground of value, a belief basic to contemporary societies. The chapter provides, then, a radical interpretation of technological society by showing that theological claims are needed for an ethics of responsibility. The argument extends the analysis of value and responsibility developed in Part II of this book.

In order to make this argument, I will examine in some detail the work of a thinker who challenges basic assumptions in ethics. In his theory of responsibility, Hans Jonas argues that previous forms of ethics are unable to address the problems raised by the technological extension of human power. Jonas further contends that previous beliefs about the ground or source of the moral life are inadequate as well. Ethics must attempt to fill the void left by the decay of traditional religious and metaphysical convictions about the source of value. Thus, Jonas' work returns us to the theme of the connection between responsibility and moral world-views explored in Part I of this book.[1] I now resume that discussion as a backdrop for reading Jonas, and also in order to provide the final formulations of the imperative of responsibility.

ETHICS AFTER THEISM AND METAPHYSICS

For most of Western ethics the problem surrounding power was to specify norms for its exercise so that one could determine when coercive action was justified, say in justifiable war, or when social arrangements were fair and just. As shown in Chapter 1, power was valued with respect to ends which might be achieved and also in terms of the intrinsic morality of acts of power. The assumption throughout this history of reflection was that human power has strict spatial and temporal limits. Human actions reached only so far into the future, at most a generation, and had effects only on others who were relatively near to me. These spatial and temporal limitations on human action have had profound import on ethics and theories of responsibility. The scope of responsibility was delimited by the extent of human power.

It is the case, of course, that we can extend in time the *meaning* of actions. Through the use of memory, tradition, and ritual action a community can remember and re-enact some ill act done to it, the glory of its conquests, or some great, even salvific deed, and thus find warrants in the past for present actions. This extension of the temporal meaning of actions has been important in systems of retribution as well as for the preservation of cultural traditions. Yet even in these cases, the temporal reach of *actions* is limited while their *meaning* can be transmitted historically. The spatial limitations on human actions have been more strict. Political power was limited to the community in which it was exercised, or, in the case of warfare, circumscribed by the parties involved in the conflict. Political boundaries were determined by the spatial reach of power. So too with individual actions. The human body provides a spatial source as well as limit for the exercise of power. Moralists have always placed restraints on accountability for the consequences of actions, especially when those consequences are beyond the scope of human foresight. While the Bible speaks of the sins of the fathers being visited on their children to the seventh generation, the spatial scope of power limited responsibility. The assumption operative in this belief about responsibility is one I have detailed in other chapters. The scope of responsibility has been tied to the connection between agent and act. If an act, and its consequences, cannot be directly or indirectly linked to an agent, then it is difficult to speak of responsibility for that action. In short, for most of human history it has been difficult to see that persons in one part of the world might be responsible for conditions of life elsewhere on the planet.

This has now changed. The reality of technological power and the increasing globalization of political, economic, and cultural power means a radical increase in human responsibility for human and planetary life. Ethics must address these matters. And yet, anyone engaged in moral inquiry faces a troubled situation in trying to address these new questions of human power. Indeed, if Jonas is right, traditional ethics simply lacks the resources to confront the moral problem of technological power. The root problem, I have argued in this book, is the equation of power and value in late-modern societies. The ground of value has shifted from the traditional belief that value is rooted in reality to the primacy of power. Put more precisely, being itself, the source of value, is now conceived in terms of power. The modern world no longer sees nature as creation or the human as created in the image of God; we no longer dwell in a universe wherein persons and things derive their value from a place in the system of being. In light of this fact, ethics must ask the question of the value of the power to create value. How is our capacity to make-good, as Charles Taylor puts it, related to seeing-good?

The first fact which confronts anyone thinking about moral value is the loss in the modern West of the idea of an objective moral order as the ground of value and the backing for principles of moral judgment. This has left modern moral inquiry with the task of trying to ground norms and principles in the general features of human actions, communicative rationality, particular historical traditions, or to show that such grounding is impossible and unhelpful. As we know from previous chapters, the problem of relativism and the objectivity of norms are at the heart of ethical debate. Generally, modern ethics has tried to meet these problems without appeal to metaphysical claims and thus with anti-realist ethical theories. Moral realism has been challenged on all fronts. The claim is that we invent morality rather than discover it.

However, matters are more complex. The main way in which modern ethics tried to ground moral values after theism has itself fallen to criticism. That grounding was in the moral subject as an agent in the world. For instance, Moses Mendelssohn, in an early essay on "Enlightenment," claimed that "I always place the definition of the human as the measure and goal of all of our endeavors and efforts, as a point on which we must direct our eyes if we do not want to lose ourselves."[2] Mendelssohn, Kant, and others saw the aspiration to universality and the person's right to respect without consideration of class or calling as basic to morality. They fashioned an account of human dignity grounded in the claim that the human is self-legislating and thus not determined by its preferences and attachments. The human is an end-in-itself, an intrinsic good, not to be used only as a means to other ends. In order to be genuinely moral, the maxims guiding our actions must be universalizable and entail respect for others. A crucial feature of these arguments was their picture of humans acting in a value-neutral time and space. Moral values and norms are not part of the fabric of the world but are human creations to serve specific needs and purposes. The unique character of the modern worldview is the proposition that humans are the only agents in the world. This outlook is anthropocentric and the value of human purposes is written onto a value neutral reality. But if that is the case, how then are we to understand the future of life on this planet in moral terms? Does it mean that the value of life is reducible to human power? A disenchanted worldview coupled with an anthropocentric theory of value seems to render modern ethics, like classical moral realism, unable to address the questions of our time.

Finally, contemporary ethics in the West faces the loss of the influence of Jewish and Christian theism on our culture in terms of the source of value. The full weight of this loss was blunted as long as some metaphysical claims about a moral order were possible, even when expressed in the form of general beliefs in historical progress, or as long as one could give a plausible account of self-legislated moral imperatives. With the seeming failure of these two strategies for justifying the moral life and the specification of the form and content of moral discourse, ethicists are once again facing the import of the end of classical theism. In classical theism, the moral order of life was grounded in the will of a personal deity, or nature manifested a moral order because its origin and end, beginning and telos, was God. The reasons for the decay of traditional theism are well known; I have recounted some of them earlier in this book. The point is that we have lost metaphysical, anthropological, and theological means for providing norms for life, for grounding moral identity, and for interpreting moral terms. With this loss, I have been arguing, power becomes the central value and the maximizing of power becomes the purpose of human existence.

What can we say about the source of value in the wake of the failure of traditional theistic and metaphysical beliefs? In answering this problem Hans Jonas reconsiders both metaphysical and theistic discourse. The importance of his work, then, is that Jonas will not allow us to assume that the project of modern ethics can continue confident that these issues have been laid to rest. He argues that in order to propose and defend an ethics we must grasp and articulate some claim about the intrinsic worth of being itself that will warrant a moral standard sufficient to guide human life and moral judgment.

What is meant by intrinsic, or non-instrumental, worth? By intrinsic worth we mean that which cannot be properly recognized for what it is – a person or object, for instance – and simultaneously be used or valued only for some other end or purpose. The recognition of intrinsic worth is the recognition of something that thwarts self-love and wanton domination by arresting one's attention and care with an objective and inescapable claim to be respected and enhanced. Intrinsic worth is the source of the sense of responsibility. Traditionally, Christians have claimed that God alone is intrinsically good and that all other things are to be enjoyed and used with respect to their relation to the divine. We are to love others and ourselves *in* God, Augustine argued. Kant claimed that the only good in itself is the good will, even as he insisted on respect for persons as ends in themselves. The idea of rational freedom evokes the *feeling* of respect. Dialogical theories of responsibility tried to articulate intrinsic worth through the I–Thou encounter. That which has non-instrumental value encounters the self as a "Thou." Emmanuel Levinas has recently argued that the face of the other utters a command which binds the self. When, as Ricoeur notes, the face of the other raises itself before me, above me, it is not an appearance that I can include within the sphere of my representations. To be sure, the other appears, his face makes him appear, but the face is not a spectacle; it is a voice. The voice tells us, "Thou shall not kill." Each face is a Sinai that prohibits murder.[3]

Traditionally, intrinsic worth was conceived in terms of God, rational freedom, response to a Thou. What is crucial about Jonas' argument is his rejection of these ways of considering intrinsic worth. Instead, he grounds value in natural purposiveness and moves ethics beyond the assumption, seen in the above examples, that intrinsic worth must be conceived of in terms of personality, be it of God, humanity, or the Thou. Jonas is concerned, as this book is, with how value is manifested to and experienced by moral agents. And he links this experience to what grounds the significance of human life and the very meaning of moral concepts. This argument requires that Jonas consider again the relation between ethics and metaphysics.

Jonas challenges long-standing assumptions about the relation that ought to obtain between ethics and claims about the nature of reality, that is, metaphysical reflection. He tries to overcome the split between ethics and metaphysics by means of what he calls the "paradox of morality." As we will see in more detail shortly, this paradox coheres with the emergence of freedom within the biosphere, a freedom defined as the self-affirmation of life against its negation. The right of finite life to exist is grounded in the purposiveness which finds moral expression in human action. What is more, the question of the role of metaphysics in ethics relates to Jonas' assessment of religion. He believes that religious ideas have lost all plausibility in the modern West. Yet because of the need in ethics for some connection between intrinsic worth and moral responsibility, Jonas bemoans the moral impotence of religious ideas in contemporary life. The challenge of his ethics is to suggest that the ethicist must think about the moral life within this religious loss, and even in terms of the "self-limitation" of God. Later I will elaborate what this concept means in his ethics and its relation to the experience of intrinsic worth and moral discourse.

Thus, Jonas accepts the dictum that ethical discourse is

logically distinct from religious beliefs and convictions. Yet he does not think that this settles the question of the entanglement of the moral life with other beliefs about the nature and value of reality, including religious beliefs. Religious beliefs add to meta-physical beliefs the problem of how we relate to the conditions of our existence. Is that relation one of confidence, hope, fear, anxiety, dread, or reverence? For a variety of reasons, Jonas rejects a theological dimension to moral inquiry. We will see that this rejection relates to his attempt to articulate the normative center of ethics around the intrinsic worth of finite reality itself, a worth that generates a moral imperative. In order to advance his argument, Jonas turns to an account of time as the horizon of moral inquiry. He examines the end of time – the threat to the future of life and thus the motive of fear – as the means by which to articulate how moral norms cohere with and yet transcend specific states of affairs. The question of the significance of the moral life within the context of the whole of things, a question previously answered through religious conviction, is addressed by examining the temporal horizon of moral behavior.

Jonas' ethics focuses then on the problem of determining an absolute moral standard after the end of traditional theistic and metaphysical warrants for ethics. I now turn to his work and focus on how Jonas addresses the problem of moral norms. There are two reasons why this is a helpful way to read his ethic. First, while it does not encompass the totality of moral inquiry, concentration on the normative and meta-ethical dimensions of ethics directs attention to the justification of moral beliefs as a central problem facing philosophers and theologians.[4] It allows us to highlight the distinctive elements of Jonas' thought and the challenge he poses to Christian ethics. Second, this reading of Jonas enables us to identify how the religious relates to the moral at a most decisive point.[5] If religious and moral reflection are interdependent, then they are most interdependent concerning normative claims and the justification of an ethics; the good or right expresses in different terms what the believer means by God or the will of God. However, if moral and religious reflection are seen as separate modes of discourse, then they are most distinct at precisely this point. This would mean, for instance, that what

God commands, or at least what the believer holds that God commands, must be tested by independent criteria of moral judgment to see if God and those commands are worthy of obedience. A focus on how Jonas addresses the problem of moral norms has, then, the virtue of concentrating our inquiry on these decisive points in ethics.

RESPONSIBILITY AND THE MORAL PARADOX

For Jonas, modern technology is an ontological event in history; it has changed the nature of human action. Technology so extends human power that the future is under our control and therefore also our responsibility. For Jonas this is the moral dilemma of our time, since in his judgment ethics, as the regulation of human action, concerns power. As he puts it, "our action is a function of our power, of that which we are able to do."[6] Because technology extends human abilities into all domains of life, it is of profound importance. "The sphere of our action," Jonas notes, "now stretches itself over the globe, perhaps to the meaning of future generations."[7] The earth, future generations, the genetic structure of various forms of life, and even history itself now fall under the domination of human action and technological efficiency. The loss of religious faith in the modern world as the grounds of the moral life, along with the change in the nature of action, creates a space of virtually unlimited human responsibility for life on this planet.[8]

Jonas begins to develop an ethics of responsibility for a technological age with a claim about freedom. As a condition for moral action, freedom is best understood as an instance of purposiveness. The basic idea of purposiveness is the self-affirmation of life against death. This is because, as Jonas writes, in "every purpose being declares itself for itself against nothingness."[9] Moral freedom emerges in the biosphere because of the capacity of human beings to differentiate themselves from their environment and to make choices about how to act. Human freedom is a unique expression of the basic fact of purposiveness; freedom introduces complexity and increased contingency into the world, but it also entails an affirmation of life. Since we now

have the power to end life on this planet, it is vitally important
that human beings take responsibility for the consequences of our
actions. Given this demand, human beings must continue to exist.
Jonas advances, then, a philosophy of emergence in which mind
and freedom have marked continuity with the rest of reality and
in which human beings bear special possibilities and burdens.
This has the effect of asserting the intrinsic worth of being as the
condition for all other goods. It also means that human freedom
now bears the weight of the self-affirmation of life. Any moral
imperative must be specified relative to the structure of free
action within the community of life.

Based on this argument, Jonas formulates an imperative of
responsibility for our age: "Do not compromise the conditions for
an indefinite continuation of humanity on earth."[10] This means
that we are "not responsible to the future human individuals but
to the *idea* of Man, which is such that it demands the presence of
its embodiment in the world."[11] The idea of Man, the idea that
humanity is an end-in-itself, is to guide action because it specifies
what ought to be embodied in the world. The idea of Man
articulates a reality, a good, that ought to be respected and
actualized regardless of what we happen to care about. If we
grasp the meaning of this idea, then we understand that it must
be embodied. This is, we can say, Jonas' ontological proof that
humanity *ought* to exist. If we understand this *idea* we do not know
that humanity exists (as in the case of the ontological proof of
God's existence), but, rather, that humanity *ought* to exist. The
proof is practical and prescriptive rather than theoretical and
descriptive in character; it concerns norms and prescriptions for
human action. The idea of Man indicates the good which ought
to be respected in all action, that is, the conditions necessary for
the continuation of human life on earth.

Granting the presence of human life *in* the biosphere conceived
of in terms of "emergence," how is the imperative that there
ought to be finite life in the future actually known in the present
so that this knowledge might move us to sacrifice present interests
for future well-being? That is, how can our self-understanding be
transformed so that we recognize that future generations of
human life make a claim on our present conduct? In order to

answer this problem, Jonas begins with how the imperative of responsibility is known. He argues that "a 'command' can issue not only from a commanding will, for instance, of a personal God or a Thou, but also from the immanent claim of a good-in-itself to its realization."[12] There are certain ends, goods to be respected, which also demand their realization. Jonas returns to Max Weber's insight that responsibility is about the future effects of present actions. Yet Jonas does not understand the politician to be the prime example of responsibility. For Jonas, the end or good which ought to be realized, Man, is paradigmatically seen in the face of the child. In the child we see what simply ought to be embodied in the world. The parent is the clue to responsibility.

In every act of procreation human beings affirm, at least implicitly, the value of continued existence. The child is the face of the human future for whom we are responsible even though future life bears no reciprocal duties to us. Just as a child does not have commensurate reciprocal obligations to the parent, so also the future claims our responsibility without itself being bound by a commensurate demand. The experience of the child allows us, again, to formulate the moral imperative: never endanger the existence of humankind. The imperative, while seemingly anthropocentric, pertains to all of life: we are responsible for all of life because human freedom and power extends in time and space within the community of life. The articulation of this norm also shows us that the biosphere is the very condition of morality and thus bears intrinsic worth.

The imperative of responsible action, as Jonas formulates it, is ontologically basic because it is about the emergence of freedom as the self-affirmation of life against its negation. Responsibility makes the perishable future of freedom and thus of life itself basic to other rights and goods. This establishes an objective, realistic norm for Jonas' ethics of responsibility. And because freedom has emerged within being, responsibility is basic to the continuance of life as the ultimate context of morality; the imperative of responsibility has universal scope. The imperative of responsibility rests, then, on the insight that for human life to exist is good. This designates the objective character of the norm of responsibility; it is not reducible to subjective wants and preferences. The moral imperative is universalized through the idea of Man; humanity itself, and not specific individuals or groups, ought to be realized.

However, an objective norm must also be related to a subjective motive to act. A moral norm is irrelevant if it does not move agents freely to live by it. (This is why the imperative of responsibility as I have formulated it had to be correlated in Chapters 7–8 to levels of value and motivation.) In order to move agents to act on the moral law, Jonas contends that being itself must be the cause and the object of reverence. We must have some recognition or sense of the goodness of being which moves us to live by the moral law. So, Jonas writes,

> Being (or instances of it), disclosed to a sight not blocked by selfishness or dimmed by dullness, may well instill reverence – and can with this affection of our feeling come to the aid of the, otherwise powerless, moral law which bids us to honor the intrinsic claim of Being.[13]

Jonas' claim is that perishable existence is the object of the feeling of responsibility. This feeling manifests the intrinsic claim of being on us in terms of what we care about. While this claim is "heard" at the level of "feeling," it can move persons to act on the moral law formulated in the imperative of responsibility. This "feeling" is not an irrational sensation. Rather, it is a sense of responsibility basic to understanding the world and human life in moral terms. And by pursuing the good, a different level of the self-affirmation of life – a higher self – comes into being as an indirect consequence of this pursuit. Responsibility is basic to authentic human being, its good, and its capabilities for autonomous moral agency.[14] Yet because it is the *object* that is to be pursued, and not the higher self itself, the "intrinsic right of the object is prior to the duty of the subject."[15]

The plausibility of Jonas' ethics depends, then, on the claim that being does in fact evoke reverence and binds us through the feeling of responsibility to live by the moral law. But a tenable ethics must acknowledge that agents are selfish and that our moral sensibility and perceptions are deeply distorted. What we care about and what we ought to respect can and do conflict. Jonas recognizes the problem. Granting the fact of selfishness, he holds that one must appeal to fear, to imagined *malum*, and not to awe or reverence as crucial to moving people to care for the future of earthly life. The threat of extinction, non-being, motivates an affirmation of life and care for the future of life. Moral motivation is born of self-interest in the face of the threat to future finite life; the threat to finite life instigates the transformation of moral understanding.

There are reasons to question this claim about fear as moral motive. Our previous examination of levels of values and the variety of motivations, such as esteem, shame, and, more basically, recognition and care, leads us to reject Jonas' analysis of moral motivation as too simplistic. But more to the point here, if the idea of Man is to determine the will because it tells us what *ought* to be realized, then this idea itself must not dull our reverence for being as basic to moral action. It must enable us, as Charles Taylor puts it, to "see-good." But the idea of Man, I judge, can do so if and only if it draws on the force of something not threatened by the negative and which discloses the worth of existence. Moral insight must be mediated by some idea, symbol, event, or name other than the idea of Man, since, on Jonas' own admission, this idea rests for its binding force on some claim about being. Jonas's argument, it would seem, requires some claim about creation and creature, that is, a properly theological claim, in order to sustain it. And yet he rejects traditional theism with respect to the demands of contemporary ethics. Simply put, Jonas, like Kant, argues that it is no longer possible to appeal to the divine as commanding or rewarding moral behavior. The idea of heteronomous commands and rewards for obedience is contrary to morality. But Jonas actually forwards a more penetrating criticism of theistic belief than Kant, a criticism basic to his own ethics.

Jonas contends that classical theism "extrapolates into an ever-present order of abstract compatibility" between God and the human good. God is eternal and so too is the human's highest good. The good is beyond time. In classical Western thought the moral life was construed as a journey into eternal presence. The moral agent was perfected, became other than its given self, and thus became more real through devotion to a transcendent good. For instance, St. Augustine contrasted the temporal "City of

Man" with the eternal "City of God" which struggles in history but will transcend history. The object of ethics was a transcendent good or God that was to be appropriated by moral action in time, while itself transcending time. As Jonas puts it, in traditional theologies "the imperishable invites participation by the perishable and elicits in it the *desire* thereto."[16] Conceived in this way, the moral life had the paradoxical character of allowing a higher self to emerge indirectly from concern for, or obedience to, this desired transcendent "object." In our age, by contrast, we must conceive of moral existence within the full range and fragility of the temporal continuance of life. The imperative of responsibility must bear on the future of freedom and hence the self-affirmation of life. Jonas contends that a plausible moral imperative in our time "extrapolates into a predictable real *future* as the open-ended dimension of our responsibility."[17] Thus, while Jonas diverges from traditional ethics because of his metaphysics of emergent freedom, he still sees this paradox of a higher self arising from duty as definitive of morality. But, given the changed nature of human action, the object of morality must be different than traditional theistic ethics even if not morality's paradoxical character.

According to Jonas, the object of moral concern and the standard of good and evil must now be the future of life and not the divine or transcendent good. When one could appeal to an eternal God, the temporal structure of human life did not bear so directly on the status of moral norms. The imperative of responsible action as Jonas formulates it is thus ontologically basic not only because it is about the emergence of freedom as the self-affirmation of life but also because it focuses moral attention on finite, temporal life. By pursuing this good, a higher self – a different level of the self-affirmation of life – comes into being as an indirect consequence of such action. Responsibility for the perishable is basic to authentic human being, its good, and its capabilities for autonomous moral agency.[18] It makes the perishable future of life itself basic to other rights and goods. The source of morality, Jonas holds, is the recognition that the standard for distinguishing good and evil is an affirmation of being over nothingness. Jonas has shifted the warrant of ethical norms from God or the self to the claim of the finite to exist. He has attempted to articulate a norm in time without appeal to the eternal being of God or an immortal soul acting amid the flux of change. The imperative of responsibility as he formulates it is rooted in the fact of purposiveness and the demands on human freedom. Jonas has presented a normative ethics by means of a principle of responsibility correlate to human power.

In Jonas' ethics of responsibility we see a complex response to the troubled situation facing moral philosophy. He provides an account of the relation of nature and freedom which gives ontological import to moral imperatives without demeaning human freedom. Similarly, his ethics provides a reformulation of realism in ethics, at least with regard to the relation of morality and beliefs about the world (metaphysics). He also introduces a fundamental alteration from traditional theistic ethics: the paradox of morality is set within the temporal structure of life which grants imperative force to the possibility of a viable future. The universalizability of this imperative is not a matter of its status as a maxim of the will (for example, the categorical imperative) but concerns the ontological conditions of life. Finally, the good as a moral concept is understood with respect to the perishability of life (the idea of Man) rather than through some idea of a transcendent good or God.

RESPONSIBILITY AND THE SELF-LIMITATION OF GOD

Thus far we have seen how Jonas attempts to justify an imperative of responsibility and to reclaim the possibility of ethics amid the criticisms of traditional thought and the pressing problems of our time. He presents a realistic ethics of responsibility in the face of the threat to a future for humanity and the earth. The focus on the future in his ethics concerns the threat to being, that is, the possible destruction of life on earth. This brings us to the relation between religious and moral claims in ethics in terms of the source of responsibility.

As we have seen, Jonas uses the language of crisis, what he calls a heuristics of fear, to move persons to act on the moral law. The imagined threat of destruction is to move us to responsible action.

The religious question emerges precisely with regard to the threat to being and thus the question of the goodness of life. It arises as an existential question, a question about the meaning and value of existence, rather than as an assertion on dogmatic grounds. Because of this, it is possible, in principle, to engage this question within the compass of moral inquiry. That is what Jonas does. He is mindful that for our culture the decay of traditional faith means not only confusion about moral norms, but also a loss of moral conviction. He too seeks to redress this loss. It is thus important to grasp how Jonas relates his moral discourse to an assessment of our religious situation. Doing so will take us one step closer to the conclusions I will draw for an integrated ethics of responsibility.

In order to enter this level of reflection we must shift the focus of our examination. With a general outline of Jonas' ethics in hand, we can now ask how he construes the relation between the mode of being of the divine and a moral norm. The phrase "mode of being" refers to the way something presents itself, if at all, to be understood. For example, one could ask, as theologians do, how God is being God in our time? Martin Luther, in his commentary on the first commandment, understood "God" to be our highest good; trust of the heart and one's god go hand in hand. But Luther also understood God's way of being God as Judge and Redeemer revealed in Jesus Christ. Stoic philosophers looked to the nature system as divine. For them, the mode of God's being God is presented through the order of nature. Currently, some theologians speak of God as liberator of the oppressed. As Beverly Harrison puts it, God takes form in the struggle for justice.[19] Thus, to ask about the mode of God's being God is to explore the way in which the divine is known and experienced. The same point could be made about ethical norms or the experience of value. The sense of responsibility, I argue, is how we experience the mode of being of the non-instrumental worth of the integrity of life.

Jonas characterizes the divine mode of being in our time as a withdrawal from our everyday experience. The God of traditional theism is simply absent from our experience of the world and the moral life. Furthermore, after the Holocaust and the horrid silence of God in its midst, Jonas argues that Jews must speak of the absence of God creating the space for human responsibility.[20] Better stated, God has limited God's power thus creating a space of human responsibility. In this space it is appropriate to appeal to fear, to imagined *malum*, and not religious awe or reverence, as crucial to moving people to care for the future. This is the religious reason for the rhetoric of crisis, the "heuristics of fear," in Jonas' ethics of responsibility. When the object of moral devotion is an eternal, transcendent good, then awe and reverence seem to qualify experience, but if the object of concern is a perishable future threatened by human power amid God's absence, fear is the emotion that moves one in devotion to the finite good.

Jonas understands the mode of being of the divine in our time as self-limitation. God does not secure a moral order, as in traditional theism, nor is the idea of God, as Kant argued, postulated on moral grounds. Our experience of the sacred is one of absence creating the space for the affirmation by persons of finite life in its radical finitude. Human beings have their moral being precisely in God's self-limitation and thus are responsible for the self-affirmation of life through freedom. The meaning of moral existence for Jonas is, then, that "life" is submitted to the risk of human freedom. God transcends the world and creates in this transcendence the moral space and problematic of human existence. In making this claim, Jonas asserts the importance of the Jewish idea of creation for our understanding of reality. Yet, while God is "transcendent," the object of responsibility is not; it is, in fact, the future of finite life on this planet. Jonas insists on the distinction between the object of responsibility and the divine, especially in the post-Holocaust era. Only God is transcendent and properly conceived as creating the space of responsibility. The reality of human self-transcendence, or what Jonas calls the moral paradox, is decidedly different than the mode of being of the divine. Moral existence is a coming into the fullness of what it means to be human in the finite world; it is the self-affirmation of life in freedom. The divine transcendence is the self-limitation of God which creates the space for freedom and the condition for binding moral claims. For Jonas, no religious experience is necessary for the principle of responsibility to remain binding.

We have a moral obligation to viable future life regardless of beliefs about or experiences of the divine.[21]

Yet Jonas is confusing at this point. If the self-limitation of God creates the space of responsibility, is not the moral paradox itself – the emergence of freedom in a higher self amid the community of life through obedience to the moral law – an opening to religious claims and experience? Jonas seems to provide the grounds for some form of moral theism, that is, that ideas about the divine are based on the dynamics of moral experience and freedom. Yet Jonas' basically Kantian argument stops short of the conclusions Kant himself drew about the idea of God. Put differently, if the transcendence of God – the divine self-limitation – is the condition for the possibility of the moral life because it creates the space of responsibility, must we not also consider the relations between God, moral existence, and claims about reality?

We have seen that Jonas specifies the norm of responsibility with respect to some claim about our experience of intrinsic worth against the threat to finite life, and, equally, with respect to how God is being God in contemporary experience. These are actually correlative claims. The experience of intrinsic worth, and thus moral obligation, is tied in Jonas' work to a claim about the self-limitation of God. Jonas poses the question of whether or not we can adequately understand any experience of moral obligation without reference to how we construe our relation to reality. Modern ethics has insisted that conditional human life is the necessary and sufficient grounds for categorical norms. To be sure, Jonas is skeptical about the capacity of traditional religious discourse to speak to this question. But he also argues that previous assumptions about the source of moral norms in human autonomy or social convention outside of reference to metaphysical claims can no longer be granted without question. I have tried to show, then, how and to what extent Jonas raises the question of "God" as the symbol for the unconditional within moral inquiry.

This returns us to Jonas's ontological proof of the idea of Man as the reality which ought to be embodied in the world. The original version of the ontological proof by St. Anselm in the

Proslogion argued that if one grasps the idea of God as that than which nothing greater can be conceived, one understands the necessity of God's existence. Since it is more perfect, maximally greater, to be than not to be, it is more perfect for something to exist in reality than simply in the intellect. If one denies the existence of that than which nothing greater can be conceived, but can think this idea, one also thinks the possibility of something greater, that is, something which necessarily exists in reality and in the intellect. And this then must be God. Debates about the validity of this proof center on whether or not existence is a predicate and also on the idea of perfection. My concern is not to enter the long debate about this proof. It is, rather, to specify how this argument attempts to articulate the unconditional element in moral consciousness as the affirmation of the goodness of being – a point Jonas also insists upon as the source of responsibility.[22] The proof does so by showing that the identity of being and goodness is basic to our consciousness of being agents.

In light of the present inquiry we must make two revisions in the proof. We must first see that the form of consciousness we are examining in an ethics of responsibility is the consciousness of our being agents, beings who make choices about how to exercise power in relation to others and our environment. While St. Anselm focused on what is believed about God, and therefore the mode of consciousness entailed in the assent of believing, our discussion is concerned with the moral meaning of faith and hence our self-understanding as agents. The connection between being and goodness which we are seeking must be basic to our experience of being agents. If this can be shown, then the power to act cannot be the sole value, and yet human efficacy will be endorsed as integral to human well-being. Second, Jonas prompts us to focus on the connection in the proof between existence and goodness. The Anselmic version of the proof moves from the idea of non-temporal perfection, goodness, to establish the necessary existence of God. But we are forced to ask about the goodness of finite existence. How can we assert that finite existence *ought* to be respected and enhanced because it *is* good?

We see the connection between these two points in Jonas'

appeal to purposiveness and thus the basic sense of human
efficacy. He argues that human beings are conscious of their
purposive actions. Moreover, in every act to bring about some
state of affairs in the world there is an implicit affirmation of the
goodness of being over non-being. To act at all is to endorse the
relation of goodness and being as basic to one's self-understanding as an agent. Jonas formulates the imperative of responsibility to express this insight in terms of a norm for choice. We
ought to affirm the possibility of continued life and thus human
responsibility on this planet. This is an imperative of responsibility because it arises from the connection between some causal
agent and the effects of action, the state of affairs brought about
by human action. But the question is whether or not this point
about purposiveness can be rendered intelligible without something like a theological claim formulated in a version of the
ontological proof. Can we make sense of the immanent claim of
a being to realization which resonates with our sense of being
agents without articulating that claim in theological terms? Is the
affirmation of being over non-being, reality over destruction,
written into the fabric of reality? And even if it is, why ought
this to determine our moral choice? Stated in terms of the proof,
can the goodness of existence endemic to our consciousness as
agents be conceived without endorsing the reality of an unconditional good, that is, God?

Here we reach the irony of Jonas' ethics of responsibility. On
the terms of his argument it is the self-limitation of God that
creates the unlimited space of human responsibility. The divine
mode of being is precisely the source of human responsibility and
with it the demand to respect and enhance finite life. The *idea* of
the self-limitation of God is the idea of a reality (God) the
existence of which creates the possibility of responsibility and thus
is the source of morality. The sense of responsibility intrinsic to
the self-understanding of agents, that is, the sense of the identity
of goodness and being manifest in the power to act, is an implicit
testimony to the reality which is the condition of finite goodness.
It is, to use traditional Christian terms, the call of conscience in
terms of the integrity of life. In every action we explicitly affirm
existence and in doing so we implicitly endorse that which creates

the conditions for responsibility. In so far as any act brings about some state of affairs, even the terrifying act of the denial of existence, one endorses in every action the reality which is the condition for the power of human action.

This insight must be formulated theistically because the idea of God, I have argued, conjoins power and value. Here is, I judge, a moral-ontological proof of God. Whether or not we explicitly sense the presence of God, every human act endorses the reality of God in the affirmation of being against its negation and fragmentation. If we think the identity of God as presented in the ways of naming God in Christian faith, then we grasp the affirmation of the reality of God in our consciousness of being agents. God is the symbol in Christian faith for the unconditional source and possibility of moral responsibility. God makes responsibility possible by creating and sustaining the moral space of life. And this means that the imperative of responsibility must be reformulated theocentrically in terms of an integrated ethics of responsibility and not, as Jonas holds, through the idea of Man. In all actions and relations we ought to respect and enhance the integrity of life *before God*. Acting on that imperative is an affirmation of the very condition of action and with it an affirmation of the being of God. The imperative of responsibility as I have formulated it articulates the central claims of Christian faith in terms of a directive for action.

The challenge to a theological ethics of responsibility is to demonstrate what this claim about God means for our self-understanding as moral agents. In previous chapters I have shown this by developing the idea of radical interpretation; we now understand the source of this practice of conscience in terms of the sense of responsibility intrinsic to the experience of being an agent. To interpret our existence and the existence of the world theologically means affirming as the principle of moral identity the imperative of responsibility. I have shown that a theological ethics of responsibility can articulate a claim which is implied in, but not made explicit by, an ethics developed to meet the challenge of a technological age. I have then specified the importance of a theological turn in the ethics of responsibility in terms of the relations of power, value, and existence.

GOD AND THE ETHICS OF RESPONSIBILITY

Given the discussion of this chapter, we must now draw some crucial conclusions for an integrated ethics of responsibility. By engaging Jonas' ethics, we can, first, elaborate further the imperative of responsibility. In the previous chapter, I gave material specification to the imperative. It can now be reformulated in two ways: in all actions and relations we are to respect and enhance *persons* before God; in all actions and relations we are to respect and enhance the *common good* before God. In light of the new global reality of increased human power, we can further specify the imperative as follows: in all actions and relations we are to respect and enhance the *community of life*.[23] This form of the imperative marks the outer reach of human responsibility correlate to the increase of human power in our time. It means that respecting and enhancing persons and the common good requires seeing the relation between those forms of responsibility and the future viability of life on this planet. Put differently, an integrated ethics of responsibility is concerned with respecting and enhancing the integral relation between persons, social existence, and life itself. It formulates a global ethics of responsibility.

These three forms of the imperative of responsibility are important. First, they give specification to the concept of the integrity of life as the good which ought to be respected and enhanced in all actions and relations. The integrity of life means the good of persons, communities, and the ecosphere as these manifest in different degrees pre-moral, social, and reflective goods. The imperative of responsibility is comprehensive or universal because all moral values can be specified with respect to these forms of the imperative. Second, these specifications of the imperative articulate the scope of the moral community. The moral community is not delimited to rational agents or even sentient beings, but extends to the full reach of finite existence before God. This has the effect of making the exercise of power subservient to respecting and enhancing a good beyond itself, the good of the integrity of life, grounded in the being of God. And, finally, in the light of the present chapter, the imperative is categorical in that its object, moral agents, and also the divine source of being and goodness are implicitly affirmed in every action and relation. By interpreting its existence through this imperative in the act of radical interpretation, the agent or community of agents endorses the condition of human life as agents. To deny the imperative of responsibility is to deny the condition of one's own existence as a relational agent and likewise to deny the reality of God. The imperative of responsibility within an integrated ethics thereby answers Jonas' concern for an ethics of life.

This brings us to the second conclusion of this chapter. I have offered an ethically modified form of the ontological proof to demonstrate the necessity of theistic discourse in ethics. How then is God to be conceived? More pointedly, how is God to be conceived and named with respect to the challenge which human power poses to ethics? Jonas, as we have seen, speaks of the divine being in ethics in terms of the self-limitation of divine power creating the space of human responsibility. Yet this is a negative claim; it makes no positive affirmations about the divine in the mediation of moral understanding. The positive affirmation Jonas does make, we should recall, centered on the idea of Man. I have argued that the idea of Man can enable us to see the good only if it draws on the moral force of something not threatened by the negative and which discloses the worth of existence. Jonas' conception of the divine self-limitation cannot serve this role precisely because it is not a conception of power, but, rather, the denial of power. But as agents we exercise power. And it is here, I judge, that the claims I have made about the divine throughout this book are germane.

In Christian ethics, ultimate power when it evokes gratitude and reverence is identified as "God" in so far as it respects and enhances finite reality. God is the name for the self-interpretation of ultimate reality in which power is transformed in relation to the worth of contingent reality. The central symbol of Christian faith is the means for the radical interpretation of our own existence as agents in the context of temporal life. By interpreting human life and the world theologically, we articulate the most basic ethical claim: value is not reducible to power and yet the capacity or power to act is a basic human good which ought to be oriented toward respecting and enhancing the integrity of life. Acting on that conviction, Christian faith claims and enacts the moral paradox. The imperative of responsibility in an integrated ethics merely formulates this conviction as a directive for action. The ethical good, the reality of moral integrity, is then nothing else than to constitute one's life and community in God.

This allows us to see, finally, the meaning of the claim that we are to respect and enhance the integrity of life *before God*. In all actions and relations we affirm the condition for responsible existence, the divine reality, as this entails interpreting life through the central symbol of Christian faith, the identity of God. The divine reality as the condition for all existence is the ground of a hermeneutically realistic ethics of responsibility. Interpreting our life through Christian beliefs about the identity of God – that "God" is the self-interpretation of ultimate power – is the means to discover the moral truth of our existence as agents. This act is the subjective correlate to the objective grounds of a Christian ethics of responsibility. Understanding the imperative of responsibility in this way means that the ultimate human good, the ethical good of moral integrity, is to be found in a life dedicated to the divine life, a moral project which does not flee worldly life to a transcendent goal, but which is a response to the goodness of existence itself. It is because of this goodness that gratitude and benevolence, not fear, move Christian life. Gratitude and benevolence are the Christian transformation of the primal moral motives of recognition and care empowering persons to respect and enhance the integrity of life.

CONCLUSION

The irony of the technological age is that with the increase of human power the goodness of existence is affirmed in human action while the policies which direct such action demean and destroy the possibility of continued existence on this planet. The challenge of our age is to put technological resources to work in respecting and enhancing the integrity of life. There is no hope nor possibility of returning to some supposedly innocent pre-technological state. Human beings exercise power in fashioning a human world. The question is how that power can and should be exercised.

The discussion of power and responsibility in this chapter has allowed us to provide further specification to the imperative of responsibility, and also to clarify its theological grounding. The force of the argument has been to show that power can and should be deployed to respect and enhance the integrity of life. And it has also been to demonstrate that, in so acting, the value of life is sensed, at least implicitly, as *before God*. Christian moral philosophy must make that affirmation explicit and show what it means for how we can and should live.

CHAPTER 9

Responsibility and Christian ethics

In this book I have argued that moral responsibility only makes sense within a set of commitments held by persons and communities about what is good, the nature of moral agents, principles of choice, and also a view of the world. From the perspective of Christian faith, I have presented a theory of value rooted in created reality, developed a theocentric imperative of responsibility, and also examined the formation of moral identity and assignments of responsibility. In presenting this account of Christian ethics, I have also claimed that a theological construal of the world finds the source of responsibility in the reality of God, rather than in the increase of power. I have shown how this insight extends the scope of moral relations to the integrity of all life. The book, like Christian faith itself, is then a radical interpretation of contemporary culture based on an understanding of life before God.[1]

The purpose of this concluding chapter is to show how the book as a whole validates its claims. In doing so, I will have established why Christian ethics must be formulated as an integrated ethics of responsibility if it is to remain consistent with its basic commitments. We can begin by clarifying the perspective we are trying to validate by drawing together the strands of the argument of the book.

PERSPECTIVE IN ETHICS

All moral reflection is undertaken from some perspective on the meaning and purpose of human life in the universe. Naturalistic ethics, for instance, defines human life in terms of our relation to natural processes. Idealistic ethics, like Kantian ethics, concentrates on the human capacity for deliberation about courses of action, and thus the ways in which thinking, feeling, and acting agents transcend the natural determinations of life. Social theories of responsibility define human life in terms of social roles and relations. Christian ethics contends that human beings live, move, and have their being in God. Our most basic relationship to the universe is a relation to the divine. The theological point of view within ethics means that the human good is found in a resolute commitment to the divine reign in all reality. We are to respect and enhance the integrity of life before God.

One task of theological ethics is then to validate this perspective within moral reflection. And in so far as the justification of an ethics must be undertaken on grounds other than simply the appeal to revelation, Christian ethics faces the same difficulty in establishing its truth as any other form of moral inquiry. In this respect, H. Richard Niebuhr rightly insisted that Christian ethics is a form of theistic moral philosophy.[2] In terms of this book, Christian ethics is a form of moral philosophy in that moral reflection is carried out through the dimensions of ethics specified in Chapter 2. These dimensions of ethics, as noted before, articulate the shape of moral understanding. The justification of an ethics is then through appeal to common moral understanding. In terms of the method of ethics there is nothing distinctive about Christian ethics. Yet Christian ethics is theistic in so far as the perspective from which life is understood, assessed, and guided is in relation to God. The goods Christians ought to seek and the rules and norms for conduct are defined in terms of relations to the divine.

However, the question of perspective in ethics is not simply about the difference among naturalistic, idealistic, social, theistic and other basic moral outlooks. No one is a theist in general. Even the god of the philosopher has been shaped by some tradition of thought and life. Each moral outlook is informed by distinctive traditions of beliefs and sustained by the life of specific communities. All discourse about God, the moral life, and the ultimate context of human existence is shaped by traditions and communities. Of course, ethics must test, criticize, and clarify a moral tradition. Ethics always travels between tradition and current life, between the beliefs, values, and forms of discourse of the past and the present concerns of life. Still, Christian moral philosophy is reflection on the moral life from a Christian perspective in order to provide meaning and guidance for how to live.

In thinking theologically about the moral life, a theologian must then examine the symbols, narratives, and ideas of the Christian tradition as resources for understanding human life and the God of reality. The theological ethical task of thinking the reality of God through the diverse literary and symbolic forms of religious discourse which render the divine identity is to articulate how claims about "who" God is transvalue power. The Christian ethicist seeks to discover the inner possibility for the exercise of power. This possibility is manifest in the symbol of creation, the idea of covenant, and beliefs about Christ's self-giving love. Yet the moral possibility of power is most centrally articulated in convictions about the divine identity, as shown in Chapters 7–8. Radical theocentric faith is the organizing principle of Christian belief and practice. This faith claims to be the means for rightly apprehending the value and meaning of our lives, the world, and others. The theological ethical task is, then, to interpret the meaning and demonstrate the truth of a faithful apprehension of existence and thereby to provide guidance for life.

A Christian perspective in ethics articulates a particular point of view for understanding and evaluating human life and the world. It provides what I called earlier a theistic moral ontology. But Christian ethics also makes a claim about the specific dynamics of the moral life. The belief is that through dedication to the divine purposes a higher, fuller form of life is enacted. The deep human aspiration to fulfillment and authenticity is answered indirectly through dedication to a specific moral project. As Jesus said: "He who finds his life will lose it, and he who loses his life for my sake will find it" (Matt. 10:39). Christian morality is built upon this paradox. This paradox redefines human fulfillment by setting life within a wider set of purposes, a more inclusive community, named the reign of God. The task of theological ethics is, then, also to articulate the meaning and demonstrate the truth of this distinct interpretation of the moral life. I have attempted to explicate the paradox basic to Christian ethics by showing how moral integrity arises in and through a commitment to respect and enhance the goods entailed in life.

Christian moral philosophy provides then a construal of the world, a moral worldview, and an account of the character of moral existence. It provides a framework for understanding and valuing life as well as a diagnosis and answer to the problem we face as moral creatures, that is, the problem of how meaningful existence is related to care and respect for others and our world. What links these elements of an ethics, I have argued, is the core insight of the Christian perspective on the moral life. The core of Christian faith is about the radical transvaluation of power in order to respect and enhance the integrity of finite life. Moral responsibility is rooted in God as value creating power, in Christ who emptied himself and took the form of a servant, and in the Spirit who empowers persons to be responsible agents (cf. Phil. 2: 1–11). In an age of increasing human power and global inter- dependence which subject life on this planet to human decision and power, Christian faith has something unique to contribute to moral reflection. It provides the means to specify the source of the sense of responsibility for finite life. And it provides the symbolic means radically to interpret moral existence and thereby direct the use of power to respect and enhance the integrity of life. If the argument of this book is correct, then our sense of responsibility is nothing less than testimony to the fact that we live, move, and have our being before God. Christian ethics renders this fact explicit, and thereby provides a way to understand the full meaning and scope of the responsible life. In doing so, theological ethics validates its own claims in terms of the dynamics of moral understanding.

Theological ethics seeks then to clarify fundamental claims of Christian thought and to show their import for contemporary existence. Given the core moral insight of Christian faith, two further convictions follow. These have been presented in this book. One conviction is that we understand ourselves rightly only in relation to the divine. The knowledge of self and knowledge of God are intimately linked. As I have put it, the primal deed of moral identity, the act of radical interpretation, is intimately bound to beliefs about the identity of God as value creating power. Understanding life from a theological perspective has the effect of binding the power we have as a self-interpreting agent to specific norms that concern the recognition, respect, and well-being of finite life. I argued in Chapters 7–8 that these norms are symbolically presented in the name of God and that we use them radically to interpret life. This constitutes the Christian conscience as a way of living in the world. The second conviction is the moral correlate of the first. Theologians have long argued that the basic moral precepts are the love of God and the love of neighbor, precepts which also entail, as St. Augustine noted, the proper love of self.[3] We are to know and love others and ourselves *in* God. In Chapters 3–5 we explored how this claim is related to the demands of responsibility: we are to respect and enhance the integrity of life in terms of its source in God. Taken together, these two claims of Christian faith mean that knowledge and love of our own and others' worth and dignity is grasped within a commitment to the transformation of power in order to respect and enhance the integrity of finite existence. The core moral insight of Christian faith is specified in terms of knowledge and love of others. Seeing ourselves and the world from this perspective necessarily commits us to the moral life. From this follows the imperative of responsibility as I have presented it throughout the chapters of this book.

JUSTIFICATION IN ETHICS

I have just outlined a Christian perspective *within* ethics and how it has been presented in this book. Yet the question arises about how the theologian or philosopher goes about validating an ethical position. Unless we are to follow blindly the dictates of leaders, parents, or our culture, it is necessary to ask about how any form of moral reflection can justify itself, demonstrate its truth. This too is a matter of method in ethics, specifically meta-ethical reflection. How is the idea of responsibility tied to the question of validating an ethics?

An ethics is validated, shown to be true, through dialectical reasoning. By dialectical reasoning I mean the art of thinking in which conundrums, questions, and problems are disclosed and resolved through questioning and answering. Our moral problems, for instance, are disclosed and resolved in terms of answering them with respect to the field of questions found in the dimensions of ethics (cf. Chapter 2). Similarly, an ethics is justified by a dialectical act of reflection. This means that Christian ethics is shown to be true not simply by appeal to scripture, revelation, tradition, or the experience of Christians. In so far as Christian ethics is a form of moral philosophy, it must face the demand to validate its claims in the arena of public moral debate.

Moral inquiry seeks to advance an ethical position, clarify basic concepts, and examine experience by engaging other ethical positions and basic moral questions. But in order to respond rightly to positions and moral problems we must meet certain demands. These demands, as Karl-Otto Apel and Jürgen Habermas note, arise out of the fact that inquiry is a communicative action. In this sense, responsibility, answering the other, is basic to communicative rationality itself.[4] What are these demands? We can note four: (1) openness to the positions of others; (2) truthfulness in the presentation of all views; (3) appeal to generally accessible evidence in making arguments; and (4) willingness to acknowledge the force of the better argument. Dialectical reasoning is the form respect for others takes in the domain of moral inquiry. It requires that we question and answer others, be accountable for arguments, and also subject the power to speak and think to criteria which respect the integrity of others and ourselves. How we think morally must be consistent with the end or purpose of ethics, which is to guide actions that respect and enhance the integrity of life.

These procedural demands on inquiry specify deontological requirements intrinsic to dialectical, communicative rationality. Yet these demands do not tell us when one ethical position is morally superior to another. In order to specify a criteria for that judgment, we must note that ethics seeks to overcome conceptual confusion, to articulate our experiences, and to isolate and to answer problems in rival positions.[5] An ethics is justified if and only if it is error-reducing with respect to comparative positions for understanding of the moral life. As Charles Taylor has argued, an ethics must bring some cognitive gain by leading from "one's interlocutor's position to one's own via some error-reducing moves, such as the clearing up of a confusion, the resolving of a contradiction, or the frank acknowledgement of what really does impinge."[6] In terms of the imperative of responsibility, a valid ethical position must enhance moral understanding.

What counts as "error-reducing" must, of course, be specified with respect to the particular moral question or position under inquiry. But three criteria of what counts as "error-reducing" are germane to all moral inquiry. An ethics is error-reducing with respect to rival positions if it achieves a greater degree of (1) clarity (2) consistency, and (3) comprehensiveness in treating the moral life than a rival account. For this reason, I have engaged other ethical positions with respect to their clarity and consistency. I have also shown that the theme of responsibility opens the whole domain of moral inquiry. Only in this way can an integrated ethics of responsibility meet the demand of comprehensiveness. Thus, a dialectical method of validating an ethics enacts within itself the demands to respect the integrity of other positions in their full complexity and also to enhance moral understanding. Moral thinking falls under the imperative of responsibility and enacts its demands.

Taylor is also correct to note that an ethics is justified only when it isolates something which really does impinge on our moral lives but which is forgotten, ignored, or denied in rival positions. An ethics is justified not simply by its internal clarity, consistency, and comprehensiveness. The validity of an ethics is also judged with respect to the claims, values, and experiences which impinge on human life and without which we cannot make sense of our lives. This is the realistic demand placed on any ethics. The most basic evidence or backing for any ethics, I judge, is a claim about moral experience and the nature of agents who can and do have that experience. In this respect, the meta-ethical question of justifying an ethics is intimately bound to its substantive claims. For instance, some forms of ethics specify moral experience in terms of obligation; the ethics is validated with respect to that experiential evidence. Other types of ethics appeal to the experience of seeking the good by moral agents. Strong, dialogical theories of responsibility examine responsiveness to others. Still other theories appeal to moral sentiments, like sympathy. These experiences become formulated in a first principle of an ethics. And this leads to the diversity of ethical positions found in the history of thought and contemporary culture. But in terms of meta-ethics, validating the ethics requires showing that moral experience must be understood in terms of that first principle.

While any ethics implicitly or explicitly is grounded in some human experience with respect to how morality is defined, appeals to experience alone will not validate an ethical position. Any human experience is ambiguous and requires interpretation. Particular experiences cannot be the transparent justification of an ethics. Appeals to experience can always be challenged with regard to the question of what really is good or right. An ethics must, accordingly, not only interpret and clarify basic moral experiences, but also provide arguments for the norms and values which are grounded in them. This enterprise is part of the dialectical enterprise of validating an ethics.

By appealing to moral experience, the theologian or philosopher tries to show what must be affirmed as the actual conditions for persons to have that kind of experience. One begins with a basic moral experience, the sense of responsibility for instance, and then attempts to uncover the conditions which must obtain in human existence for persons to have such experiences. Some of these conditions are transcendental, that is, conditions of the possibility of having an experience, and they are necessary in so far as their denial is self-contradictory. Other conditions include social, historical, linguistic, economic, and political factors which shape experience, that is, the actual situations in which persons live and know their world. But, based on the analysis of the transcendental and socio-historical conditions for moral experience, the ethicist then attempts to articulate the coherence of the moral life in the light of this new perspective on moral experience. The ethics is simply the systematic presentation of the moral life around a distinctive perspective. In order to deny the ethics, we must then be willing to deny the experience we actually have, or the way in which it has been interpreted.

I tried to show in Chapters 7 and 8 that the denial of what the symbol God means morally leaves us unable to make sense of the experience of "seeing-good" or the feeling of the claim of any being upon us to be realized in our self-understanding as agents in relation to others and the world. We also cannot understand the form of moral consciousness we have since it has been shaped by the moral tradition of the West. Thus, what does impinge, I argued, is the experience of the confluence of being and value basic to our consciousness of being agents. This is expressed in a basic commitment, an act of faith, which is constitutive of responsible existence: power can and must serve to protect and promote the value of finite life. That commitment is further formulated in the imperative of responsibility and thus the project of respecting and enhancing the integrity of life in order to secure the value of finite existence before God. The denial of this experience of value and the commitment it involves is self-refuting with respect to our intuitions about responsibility and the moral requirements on the exercise of power. This is the way I have attempted to validate theological claims within ethics.

We can say, then, that an ethics is justified if and only if three conditions are met: (1) it meets procedural demands entailed in the act of communication; (2) it articulates an account of the moral life which clarifies confusions and oversights in rival ethical positions; and (3) it specifies what necessarily impinges on human experience as a condition for us having the experiences we actually have. The strategy of justification undertaken in this book is dialectical in that it must be undertaken by engaging other positions and thus it is always open to counter-arguments. It is an ongoing process of attempting to show the greater adequacy of a position with respect to available options in ethics. Adopting this approach to inquiry within theological ethics means that an ethics is developed not simply by appeal to divine revelation, the Bible, or Christian tradition. A variety of sources must be used to develop an ethics, even as the theologian can only justify moral claims by exploring other modes of moral reflection. Theological ethics must show that it can articulate the meaning and demonstrate the truth of Christian faith in the context of ongoing inquiry into the moral life and the problems and possibilities of life.

It is important to see that this approach to meta-ethical inquiry is demanded by our moral situation and by Christian faith. In Chapter 1, I argued that the most glaring feature of contemporary technological, pluralistic societies is the widespread confusion and uncertainty about which values and norms ought to guide human action. The confusion arises out of the sheer diversity of moral outlooks in these societies, as well as the criticism of traditional beliefs which characterizes much modern life. The confusion and uncertainty about moral values and norms also arises due to the radical increase in human power in our time, which poses unprecedented questions for ethics. The approach to moral justification outlined above is meant to address these matters. By engaging other moral theories, as we did in Chapter 4 and elsewhere, we clarify the options available in ethics and also demonstrate which ethics is most adequate for answering our moral questions. This is to address the problem of moral confusion. The task of trying to isolate what does impinge on moral experience, as in Chapters 7–8, aims to answer moral uncertainty. It is to show within the experience of being an agent a sense of existence as good that is the ground for judgments about what we ought to be and to do. The approach to validating an ethics adopted in this book is undertaken then in order to address pervasive moral problems within the social context in which the question of responsibility is being debated.

The other reason for approaching the task of moral thinking in this fashion is explicitly theological. Christian faith entails fundamental convictions about the reality of God. As I argued in the previous chapter, Christians believe, first, that God is the sole necessary condition for all of reality. God is that in whom we live, move, and have our being (Acts 17:28). As Thomas Aquinas puts it:

It must be said that everything, that in any way is, is from God ...
Therefore all beings other than God are not their own being, but are
beings by participation. Therefore, it must be that all things which are
diversified by the diverse participation of being, so as to be more or less
perfect, are caused by one First Being, Who possesses being most
perfectly.[7]

Of course, classical theism, as I have shown, has come under
criticism in the modern world. But the point is that Christians
affirm that the divine is the necessary condition for all that exists.
Dialectical thinking, we have seen, seeks to isolate what necessarily does impinge on life; in a theological context this means
exploring the human relation to the divine. It aims to show that
within moral experience there is an encounter which is the source
of the sense of responsibility.

A second fundamental claim of Christian faith is that God has
acted and is acting in history. The very identity of the divine, the
revelation of who God is in relation to human beings and the
world, is known in a history of redemption, the history of Israel,
Christ, and the Church. Here too there are problems with
traditional Christian belief, since, as we saw in Chapters 1–2,
there are criticisms of any attempt to use categories drawn from
reflection on human agency to identify the divine. The point
remains, however. Christian faith entails the conviction that who
God is is revealed within the travail of history. Moral inquiry is
also historical in this sense. It seeks to chart the development of
concepts and values basic to our sense of what life is about by
engaging other positions. It affirms history as the arena in which
the truth of human existence and the good is to be revealed,
recognized, and identified. In terms of this book, I have tried to
show the importance of Christian convictions about the identity
of God for any truly radical interpretation of life which gives
moral coherence to life through time.

A dialectical approach to justification in ethics is then consistent with basic Christian convictions about God. This is the case
in so far as the approach to moral thinking outlined here seeks to
isolate conditions for the moral life and also the values which
have historically shaped moral identity and are present within
contemporary moral debate. Theological ethics must show how

Christian discourse aids in articulating, assessing, guiding, and understanding the moral life. It is to provide a moral hermeneutic of life. Throughout, I have then presented the moral meaning of Christian faith and demonstrated its import for the ethics of responsibility.

RESPONSIBILITY AND MORAL INTEGRITY

I have now clarified my strategy of justification on responsibility and Christian ethics. This has allowed me to draw together the strands of the argument around the idea of responsibility and to show the coherence of an integrated ethics of responsibility. This returns us to the fact that responsibility does not play the same role in this theological ethics as it did in dialogical theories of responsibility which dominated so much twentieth-century Christian thought. In order to bring this book to a conclusion, we must state why this is the case.

"Responsibility" is not the first principle of ethics because we can always ask: what makes encounter with others, or the Other, right and good? And if we try to answer that question, then there must be some good beyond the mere event of encounter which ought to be respected and enhanced. That good is the integrity of values, and, ultimately, moral integrity itself. But this means that the root moral phenomenon is not only being addressed or encountered by the "other." It is also the restless quest for wholeness by ourselves, within our relations to others, and, finally, by life in which we encounter others and seek to enhance life. Yet the wholeness of life is never achieved within the travail of actual life. The plight of human existence is that we are barred from this good. Persons are compelled to ask about how to live given the fragmentariness of existence. "High religion," Reinhold Niebuhr noted, "is distinguished from the religion of both the primitive and ultra-moderns by its effort to bring the whole of reality and existence into some system of coherence. The primitives, on the other hand, are satisfied by some limited cosmos, and the moderns by a superficial one."[8] Christian ethics, as we have seen, seeks to understand the coherence or integrity of all of life before God.

This problem of how to understand the integrity of reality amid the fragmentariness of existence explains why authenticity and fulfillment are the primary values of an overly secularized, technological culture. Authenticity is the quest for coherence within personal life; the search for fulfillment tries to complete life, make a person whole. And because these are quests or projects, power becomes central in life. It is only through maximizing power that the values of authenticity and fulfillment can be attained. Power thus has become the first principle of life. This is the basic faith of the late-modern world. But it is a superficial moral worldview. The quest for fulfillment and authenticity in terms of maximizing power leads to further fragmentation and frustration. It pits life against life in a circle of fear. The basic problem, as I put it before, is how to understand the power to act as a genuine human good without making it the good of life. To do so requires a power which transvalues power as the very source of morality. And this is what Christians mean by God and why they pray that the divine reign should come.

I have tried to show that responsible existence from a Christian perspective is concerned with the integration of diverse goods in human life. Similarly, I have argued that the lives of persons and communities are dependent on identity-conferring commitments. The ideas of authenticity, or truthfulness to some basic commitment, and of human wholeness, or fulfillment, are basic to a Christian ethics of responsibility. Yet I have not argued that moral integrity can be the direct purpose of human life. The paradox of morality from a Christian point of view is, again, that we cannot directly aim at the final good of human life for ourselves. Genuine moral integrity is an indirect consequence of first seeking to respect and enhance the integrity of all life before God. It is in this sense that persons must lose themselves in order to find themselves. The true integrity of existence is only received when a person's or community's life has a purpose other than its own quest for authentic fulfillment.

This insight has important implications for ethics, implications which address the current criticisms of the Christian moral vision. First, this insight means that the power to act is always to be put in the service of the integrity of life. Christians claim that this is in fact the inner truth of ultimate reality. Ultimate power, God, is known in terms of its relation to finite existence as the source and goal of life itself. Theological ethics severs the equation of power and value and insists that power serve a value beyond itself, the good of existence. This is the revolutionary nature of Christian faith, for it recasts the scale of values which determines how we see, know, and experience the world and human existence. A Christian account of the moral life does not mutilate human goods, as the critics hold. Rather, it curtails the valorization of power and thereby protects the values which permeate finite life.

Second, the paradox of morality from a Christian perspective is always a matter of faith. Living responsibly requires a profound trust that respecting and enhancing the integrity of life enacts the true integrity of an agent's own life. Christian identity is conferred by the commitment to undertake the risk of the moral life in faithfulness to the God of reality. It provides a confidence to live and act amid the fragmentariness of life beyond the superficial answers of the culture of fulfillment and authenticity. In this respect, Christian ethics seeks to counter the forms of false faith, the deceptions and illusions about human well-being circulating in a society, which seek to deny the genuine risk, tragedy, and also possibility of the moral life. In doing so, Christian ethics provides the means to address honestly and without illusion the pressing problem of power which is now at the center of life on this planet.

If my argument is right, then Christian faith actually articulates and interprets the basic experience of responsibility. That experience is the sense that contingent existence is good in its finitude and ought to be respected and enhanced. It is the experience of goodness shining through the fragmentariness and travail of existence, the awareness that being as being is good. This experience is basic to any other moral commitments we hold as well as crucial for the responsible use of power. Christian faith articulates this experience in terms of the being of God. The reality of ultimate power is good, but this selfsame power graciously respects and enhances the integrity of finite life. To understand life from this perspective is to endorse the power to act as good and yet to transvalue power and direct it to the ends of the integrity of life.

An integrated ethics of responsibility examines the transvaluation of power through the experience that it is good to exist. This basic moral experience, Christian faith claims, is a testimony to the fact that we live, move, and have our being in God. The purpose of an ethics of responsibility is to make sense of that claim and thus to clarify the moral meaning of Christian faith.

CONCLUSION

This book has had several purposes. First, I have sought to engage contemporary thought on the theme of responsibility. I have also tried, second, to clarify previous Christian accounts of responsibility and to isolate their insights and shortcomings with respect to the current demands on ethics. The third and primary aim of the book has been to outline an ethics of responsibility which articulates the meaning of Christian faith, answers basic moral questions, and contributes to contemporary moral inquiry. In this respect, I have proposed new directions for Christian reflection on responsibility. Finally, I have sought to reclaim for our time the idea of Christian moral philosophy. This form of reflection is faith seeking moral understanding; *fides quaerens intellectum moralem*. Thus what I have tried to show, and what, in fact, every theological ethics must show, is that the crucial matters of life are matters of faith, matters of identity-conferring commitments. It is to that point that Christian moral philosophy can and must and may speak its distinctive word.

Notes

INTRODUCTION

1 On agentic-relational accounts of human beings in Christian ethics see Mark Ellingsen, *The Cutting Edge: How Churches Speak on Social Issues* (Grand Rapids: William B. Eerdmans Publishing Company, 1993).
2 On hermeneutics and theological ethics see William Schweiker, *Mimetic Reflections: A Study in Hermeneutics, Theology, and Ethics* (New York: Fordham University Press, 1990).
3 H. Richard Niebuhr, *The Responsible Self: An Essay in Christian Moral Philosophy*, with an introduction by James M. Gustafson (New York: Harper and Row, 1963).

I RESPONSIBILITY AND MORAL CONFUSION

1 William K. Frankena, *Thinking about Morality* (Ann Arbor: The University of Michigan Press, 1980), p. 26.
2 On this see Paul Ramsey, *Basic Christian Ethics* (New York: Charles Scribner's Sons, 1950). Also see James Johnson and David Smith (eds.), *Love and Society: Essays in the Ethics of Paul Ramsey* (Missoula: American Academy of Religion, 1974).
3 See Paul Ricoeur, "The Golden Rule: Exegetical and Theological Perplexities," in *New Testament Studies*, 36 (1990), 392–397.
4 See Bernard Williams, *Ethics and the Limits of Philosophy* (Cambridge: Harvard University Press, 1985).
5 Sarah Lucia Hoagland, *Lesbian Ethics: Toward New Values* (Palo Alto: Institute of Lesbian Studies, 1988), p. 82.
6 See Paulo Freire, *The Pedagogy of the Oppressed* (New York: Continuum, 1970).
7 See Charles Taylor, *The Ethics of Authenticity* (Cambridge: Harvard University Press, 1992). The same point is made by the Lesbian ethicist Sarah Lucia Hoagland. She writes that "I have always regarded morality, ideally, as a system whose aim is, not to control

228

individuals, but to make possible, to encourage and enable, individual development." See her *Lesbian Ethics*, p. 285. What is one to make of agreement on this point between such diverse thinkers?

8 James M. Gustafson, *Ethics from a Theocentric Perspective* 2 vols., (University of Chicago Press, 1981, 1984).

9 Schubert Ogden, *Is There Only One True Religion or Are There Many?* (Dallas: Southern Methodist University Press, 1992), p. 48.

10 See Marion Smiley, *Moral Responsibility and the Boundaries of Community: Power and Accountability from a Pragmatic Point of View* (University of Chicago Press, 1992).

11 On this see René Girard, *Violence and the Sacred*, translated by Patrick Gregory (Baltimore: Johns Hopkins University Press, 1977).

12 See Peter A. French, *Responsibility Matters* (Lawrence: University Press of Kansas, 1992).

13 Wilhelm Weischedel, *Das Wesen der Verantwortung: Ein Versuch*, Dritte unveränderte Auflage (Frankfurt: Vittorio Klostermann, 1972), p. 19.

14 On pluralism see Paul Knitter, *No Other Name: A Critical Review of Christian Attitudes Toward the World Religions* (Maryknoll: Orbis Books, 1985); David Tracy, *Plurality and Ambiguity: Hermeneutics, Religion, Hope* (San Francisco: Harper & Row, 1987); Langdon Gilkey, *Society and the Sacred: Towards a Theology for a Culture in Decline* (New York: Crossroads, 1981); William Schweiker and Per M. Anderson (eds.), *Worldviews and Warrants: Plurality and Authority in Theology* (Landham: University Press of America, 1987); and, of course, Ernst Troeltsch, *The Social Teachings of the Christian Churches*, 2 vols., in The Library of Theological Ethics (Louisville: John Knox/Westminster Press, 1992).

15 Jeffrey Stout, *Ethics after Babel: The Languages of Morals and their Discontents* (Boston: Beacon Press, 1988).

16 Mary Midgley, *Can't We Make Moral Judgments?* (New York: St. Martin's Press, 1992), p. 86.

17 Dorothy Emmet, *Rules, Roles and Relations* (New York: St. Martin's Press, 1967), p. 91.

18 Richard Rorty, in advocating this position, says that a liberal cannot answer the question "why not be cruel?" other than by appealing to the discourse of some community that thinks it wrong to be cruel to others. See Richard Rorty, *Contingency, Irony, and Solidarity* (Cambridge University Press, 1989).

19 Emmet, *Rules, Roles and Relations*, pp. 89–108.

20 Patricia Derian, "Reconciling Human Rights and U.S. Security Interests in Asia." Hearings before the subcommittees on Asian and Pacific Affairs and on Human Rights and International Organizations of the Committee on Foreign Affairs, House of Representatives, 1982, p. 483. This passage is cited in David Little "The Nature

and Basis of Human Rights," in Gene Outka and John P. Reeder, Jr. (eds.), *Prospects for a Common Morality* (Princeton University Press, 1993), pp. 73–92.

21 See Paul Ramsey, *Fabricated Man: The Ethics of Genetic Control* (New Haven: Yale University Press, 1970).

22 See Christine Firer Hinze, "Power in Christian Ethics: Resources and Frontiers for Scholarly Exploration" in *The Annual of the Society of Christian Ethics* (Washington: Georgetown University Press, 1992), pp. 277–290.

23 Beverly Harrison, *Making the Connections: Essays in Feminist Social Ethics* (Boston: Beacon Press, 1985), p. 290 n. 5. As Larry Rasmussen notes, "power in its most elemental, vital sense is simply the 'power to,' the power of agency, human and non-human agency alike, in the vast web of this miracle called life, itself an expression of God's life." Larry L. Rasmussen, "Power Analysis: A Neglected Agenda in Christian Ethics," in *The Annual of the Society of Christian Ethics* (Washington: Georgetown University Press, 1991), p. 9.

24 Alasdair MacIntyre, *After Virtue: A Study in Moral Theory* (University of Notre Dame Press, 1981).

2 A NEW ETHICS OF RESPONSIBILITY

1 See Hans Küng, *Global Responsibility: In Search of a New Global Ethic*, translated by John Bowden, (New York: Crossroads, 1991).

2 On the idea of integrity see Mark S. Halfon, *Integrity: A Philosophical Analysis* (Philadelphia: Temple University Press, 1989); James Gutman, "Integrity as a Standard of Evaluation" in *Journal of Philosophy*, 42 (1945), 210–216; Lynne McFall, "Integrity," in John Deigh (ed.), *Ethics and Personality: Essays in Moral Psychology* (University of Chicago Press, 1992), pp. 79–94; Gabriele Taylor, "Integrity," in *Proceedings of the Aristotelian Society*, 55 (1981), 143–159; and Bernard Williams "A Critique of Utilitarianism," in J. J. C. Smart and Bernard Williams (eds.), *Utilitarianism: For and Against* (Cambridge University Press, 1973).

3 Dorothy Emmet, *The Moral Prism* (London: Macmillan Press, 1979), p. 7.

4 In a similar way James M. Gustafson has spoken of base points, or points of reference, for any systematic theological ethics. For Gustafson these include (a) an understanding and interpretation of God; (b) an interpretation of human experience, community, and natural events; (c) claims about what it means to be a moral agent; and (d) some claims about how persons *ought* to choose and judge their actions. See his *Protestant and Roman Catholic Ethics: Prospects for*

Rapprochement (The University of Chicago Press, 1978). By examining the dimensions of ethics I am attempting to provide an interpretation of moral experience by exploring the questions we ask and their relation to moral inquiry. Characterizing the dynamics of ethics is at the same time a depiction of the shape of the moral life. We can sketch the outlines of what I will explore as follows:

The Dimensions of Ethics

Interpretative dimension
"What is going on?"
|

Practical dimension —— *Meta-ethical dimension* —— *Fundamental dimension*
"What are we to   "How do we justify   "What does it mean
be and to do?"   moral claims?"   to be an agent?"

|
Normative Dimension
"What is the norm for how to live?"

I want to show now that the dynamics of moral understanding are at the same time a depiction of the structure of ethics.

5 J. R. Lucas, *Responsibility* (Oxford: Clarendon Press, 1993), p. 9.
6 On the idea of moral ontology see Charles Taylor, *Sources of the Self: The Making of Modern Identity* (Cambridge: Harvard University Press, 1990). Also see William Schweiker, "The Good and Moral Identity: A Theological Ethical Response to Charles Taylor's *Sources of the Self*," in *The Journal of Religion*, 72: 4 (1992), 560–272.
7 Albert R. Jonsen, *Responsibility in Modern Religious Ethics* (Washington: Corpus Books, 1968), p. 175.
8 Richard Gula, *What Are They Saying About Moral Norms?* (New York: Paulist Press, 1982), p. 5.
9 Max Weber, "Politik als Beruf," in *Gesammelte politische Schriften*, herausgegeben von Johannes Winckelmann, 4th ed. (Tübingen: J. C. B. Mohr (Paul Siebeck), 1980), pp. 505–560.
10 Wolfgang Huber, "Towards an Ethics of Responsibility," in *The Journal of Religion*, 73: 4 (1993), 579.
11 On naming and thinking God see Paul Ricoeur, "Naming God," in *Union Seminary Quarterly Review*, 34 (1979), 213–230 and David Tracy, "Literary Theory and the Return of the Forms of Naming and Thinking God in Theology," in *The Journal of Religion*, 74: 3 (1994), 302–319.
12 On a relational theory of value see H. Richard Niebuhr, "The Center of Value" in *Radical Monotheism and Western Culture* (New York: Harper and Bros., 1960), pp. 100–113. Also see James M. Gustafson, *A Sense of the Divine: The Natural Environment from a Theocentric Perspective* (Cleveland: The Pilgrim Press, 1994).

232 *Notes to pages 48–63*

13 Dorothy Emmet, *The Moral Prism*, p. 63.
14 See Donald Wiebe, *The Irony of Theology and the Nature of Religious Thought* (Montreal and Kingston: McGill-Queen's University Press, 1991).
15 Paul Ricoeur, *Interpretation Theory: Discourse and the Surplus of Meaning* (Fort Worth: Texas Christian University Press, 1976).
16 See Michel Foucault, *The History of Sexuality*, I: *An Introduction*, translated by Robert Hurley (New York: Pantheon, 1978). For a different position see Thomas E. Wartenberg, *The Forms of Power: From Domination to Transformation* (Philadelphia: Temple University Press, 1990).

3 THE IDEA OF RESPONSIBILITY

1 H. Richard Niebuhr, *The Responsible Self: An Essay in Christian Moral Philosophy* (New York: Harper and Row, 1963), p. 56.
2 Dietrich Bonhoeffer, *Ethics*, edited by Eberhard Bethge (New York: Collier Books, 1986), p. 224.
3 Dorothy Emmet, *Rules, Roles and Relations* (New York: St. Martin's Press, 1967), pp. 13–15.
4 Albert R. Jonsen, *Responsibility in Modern Religious Ethics* (Washington: Corpus Books, 1968), p. 3. Also see Richard McKeon, "The Development and Significance of the Concept of Responsibility," in his *Freedom and History and Other Essays* (The University of Chicago Press, 1990).
5 See Jacques Henriot, "Responsabilité" in *Encyclopédie Philosophique Universelle II Les Notions Philosophiques Doctinionaire*, edited by Sylvain Auroux (Paris: Presses Universitaires de France, 1989), II, pp. 2250–2253.
6 F. H. Bradley, *Ethical Studies* (Oxford University Press, 1988).
7 L. Lévy-Bruhl, *L'Idée de Responsibilité* (Paris: Librairie Hachette et Cie, 1844), p. 3. Also see Eric Mount, Jr., *Conscience and Responsibility* (Richmond: John Knox Press, 1969).
8 André Lelande (ed.), *Vocabulaire technique et critique de la Philosophie*, 6th edn. (Paris, 1951), pp. 426–427. This passage is cited in Jonsen, *Responsibility in Modern Religious Ethics*. Jonsen also notes that "responsibility" does not appear in the *Dictionnaire des Sciences Philosophiques* (1851) or the *Wörterbuch der philosophischen Begriffe und Ausdrücke* (1899). It did appear in English in the *Dictionary of Philosophy and Psychology* (1902).
9 Martin Buber, *I and Thou* (Edinburgh: T. and T. Clark, 1937).
10 Emil Brunner, *Truth as Encounter* (Philadelphia: The Westminster Press, 1964), p. 19.

11 Amitai Etzioni, *The Moral Dimension: Toward a New Economics* (New York: The Free Press, 1988), p. 9.

12 Jean-Paul Sartre, *Being and Nothingness: An Essay on Phenomenological Ontology*, translated by Hazel Barnes (New York: Philosophical Library, 1956), p. 707.

13 A. W. H. Adkins correctly notes that the Greeks lacked a concept of responsibility because their ethics does not center on the idea of duty. See his *Merit and Responsibility* (Oxford University Press, 1960).

14 Aristotle, *Nicomachean Ethics*, in Richard McKeon (ed.), *Introduction to Aristotle* (New York: The Modern Library, 1947), Book II, 1106b35–1107a10. Citations of Aristotle's work will be given by book, chapter, and line number.

15 Marion Smiley, *Moral Responsibility and the Boundaries of Community: Power and Accountability from a Pragmatic Point of View* (The University of Chicago Press, 1992), p. 55.

16 Aristotle, *Nicomachean Ethics*, Book III. 1. 111a21–25.

17 J. R. Lucas, *Responsibility* (Oxford: Clarendon Press, 1993), p. 275.

18 Aristotle, *Nichomachean Ethics*, Book I. 7. 1098a 15–20.

19 See Thomas Aquinas, *Summa Theologiae*, 5 vols., translated by the English Dominican Province (Westminster: Christian Classics, 1981), I/II, q. 62.

20 Ibid., I-II, q. 6. art. 1.

21 St. Augustine, *On Free Choice of the Will* (Indianapolis: Bobbs-Merrill, 1964), p. 93.

22 Judith Shklar, *Legalism* (Cambridge: Harvard University Press, 1964), p. 45. Also see Elizabeth Anscombe, "Modern Moral Philosophy," in *Philosophy*, 33 (1958), 1–19.

23 Smiley, *Moral Responsibility and the Boundaries of Community*, p. 10.

24 Hans Jonas, "Contemporary Problems in Ethics from a Jewish Perspective," in his *Philosophical Essays: From Ancient Creed to Technological Man* (Englewood Cliffs: Prentice-Hall, 1974), p. 172.

25 On the question of rights see Alan Gewirth, "Common Morality and the Community of Rights" and David Little, "The Nature and Basis of Human Rights," in Gene Outka and John P. Reeder, Jr. (eds.), *Prospects for a Common Morality* (Princeton University Press, 1993), pp. 29–52; 73–92.

26 Roman Ingarden, *Man and Value*, translated by Arthur Szylewicz (Washington: The Catholic University of America Press, 1983), p. 54. Ingarden's discussion of responsibility first appeared as *über die Verantwortung: Ihre ontischen Fundamente* (Stuttgart: Reclam, 1970).

27 The structure of responsibility is reflected in H. Richard Niebuhr's well-known definition: "the idea of an agent's action as a response to an action upon him in accordance with his interpretation of the

latter action and with his expectation of a response to his response; and all of this is in a continuing community of agents." H. Richard Niebuhr, *The Responsible Self*, p. 65.

28 William Kneale, "The Responsibility of Criminals," in James Rachels (ed.), *Moral Problems: A Collection of Philosophical Essays* (New York: Harper and Row, 1971), p. 172.

29 Also see Georg Picht, "Der Begriff der Verantwortung," in his *Wahrheit, Vernunft, Verantwortung: Philosophische Studien* (Stuttgart: Ernst Klett, 1969), pp. 318–342.

30 The language of conscience was originally meant to designate this fact. See, for instance, Paul Tillich, *Morality and Beyond* (New York: Harper and Row, 1963) and Martin Heidegger, *Being and Time*, translated by John Macquarrie and Edward Robinson (New York: Harper and Row, 1962).

31 On this see H. L. A. Hart, "The Ascription of Responsibility and Rights," in Antony Flew (ed.), *Essays on Logic and Language* (Oxford: Basil Blackwell, 1951), pp. 145–166. For a discussion of the narrative ascription of identity see Paul Ricoeur, *Soi-même comme un autre* (Paris: Éditions du Seuil, 1990), pp. 137–198.

4 THEORIES OF RESPONSIBILITY

1 Immanuel Kant, *Fundamental Principles of the Metaphysics of Morals*, translated by Thomas K. Abbot (New York: Liberal Arts Press, 1949), p. 54.

2 Ibid., p. 31.

3 Ibid., p. 46.

4 See Alan Donagan, "Common Morality and Kant's Enlightenment Project," in Gene Outka and John P. Reeder, Jr. (eds.), *Prospects for a Common Morality*, (Princeton University Press, 1993), pp. 53–72.

5 Kant, *Fundamental Principles of the Metaphysics of Morals*, p. 5.

6 Ibid., p. 18.

7 Ibid., p. 19, n. 3.

8 Paul Tillich, *Morality and Beyond* (New York: Harper Torchbook, 1963), p. 19.

9 Ibid., p. 21.

10 Ibid., p. 24.

11 Ibid., p. 64.

12 Ibid., p. 21.

13 Ibid., p. 40.

14 Dietz Lange, *Ethik in evangelischer Perspektive* (Göttingen: Vandenhoeck and Ruprecht, 1992), p. 175.

15 James M. Gustafson has made this point in his *Protestant and Roman*

Catholic Ethics: Prospects for Rapprochement (University of Chicago Press, 1978), p. 42.

16 Peter A. French, *Responsibility Matters* (Lawrence: University Press of Kansas, 1992), p. 61.

17 Ibid., p. 5.

18 Marion Smiley, *Moral Responsibility and the Boundaries of Community: Power and Accountability from a Pragmatic Point of View* (University of Chicago Press, 1992), p. 39.

19 Ibid., p. 14.

20 French, *Responsibility Matters*, p. 81.

21 Smiley, *Moral Responsibility and the Boundaries of Community*, p. 167.

22 Ibid., p. 172.

23 See Bernard Williams, "A Critique of Utilitarianism," in J. J. C. Smart and Bernard Williams (eds.), *Utilitarianism: For and Against* (Cambridge University Press, 1973).

24 Smiley, *Moral Responsibility and the Boundaries of Community*, p. 14.

25 French, *Responsibility Matters*, p. 67.

26 See Stanley Hauerwas, *Against the Nations* (Minneapolis: Winston, 1985).

27 Alasdair MacIntyre, *Three Rival Versions of Moral Enquiry: Encyclopedia, Genealogy, Tradition* (University of Notre Dame Press, 1992). On the relation of narratives, skills, practices, and truth see Alasdair MacIntyre, *After Virtue: A Study in Moral Theory* (University of Notre Dame Press, 1981).

28 Stanley Hauerwas, "The Moral Authority of Scripture: The Politics and Ethics of Remembering," in Charles Curran and Richard McCormick (eds.), *Readings in Moral Theology*, No. 4 (New York: Paulist Press, 1984), p. 243. Also see Stanley Hauerwas, *A Community of Character* (University of Notre Dame Press, 1981) and his *Vision and Virtue* (University of Notre Dame Press, 1974).

29 Ibid., p. 243.

30 Ibid., pp. 244–245.

31 Eric Auerbach, *Mimesis: The Representation of Reality in Western Literature*, translated by Willard R. Trask (Princeton University Press, 1953), p. 48.

32 Hauerwas, "The Moral Authority of Scripture," p. 245.

33 Ibid., p. 254.

34 See Emmanuel Levinas, *Totality and Infinity: An Essay on Exteriority*, translated by A. Lingis (Pittsburgh: Duquesne University, 1969). For a helpful comparison of Levinas and Karl Barth see Steven G. Smith, *The Argument to the Other: Reason Beyond Reason in the Thought of Karl Barth and Emmanuel Levinas* (Chico: Scholars Press, 1983).

35 The validity of such a position was first questioned in the Platonic

dialogue *Euthyphro*. Euthyphro tells Socrates that piety demands obedience to the command of the gods. The Socratic claim is that there is a principle of Good higher than the will of the gods. Divine command ethics is enjoying renewed interest among philosophers and theologians. See Philip Quinn, "The Recent Revival of Divine Command Ethics," in *Philosophy and Phenomenological Research*, 50, supplement (Fall 1990): 345–365 and William Schweiker, "Divine Command Ethics and the Otherness of God," in O. F. Summerell (ed.), *The Otherness of God* (forthcoming).

36　Karl Barth, "Das Problem der Ethik in der Gegenwart" (1922) in Karl Barth, *Vorträge und kleinere Arbeiten (1922–1925)*, herausgegeben von Holger Finze (Zürich: Theologischer Verlag, 1990), p. 102.

37　Karl Barth, *Church Dogmatics*, II/2, edited by G. W. Bromiley and T. F. Torrance (Edinburgh: T. and T. Clark, 1957), p. 674.

38　Karl Barth, *Church Dogmatics*, I/1, translated by G. T. Thomson (Edinburgh: T. and T. Clark, 1936), p. 426.

39　Ibid., p. 553.

40　In his practical ethics, Barth does draw on the resources of various disciplines and descriptions of human character and conduct, but from within the spheres of life constituted by the revelation of God as creator, redeemer, and reconciler. Other disciplines can be used only to the extent that they do not (1) define the good or (2) constrict the freedom of God.

41　H. Richard Niebuhr, *The Responsible Self: An Essay in Christian Moral Philosophy*, introduction by James M. Gustafson (New York: Harper and Row, 1963), p. 46. Also see William Schweiker, "Radical Interpretation and Moral Responsibility: A Proposal for Theological Ethics," in *The Journal of Religion*, 73:4 (1993), 613–637.

42　Ibid., p. 48.

43　Ibid., p. 65.

44　Ibid., p. 71.

45　Ibid., pp. 61–62.

46　Ibid., p. 64.

47　Ibid., p. 65.

48　Ibid., p. 145.

49　Ibid., p. 126.

50　H. Richard Niebuhr, "The Center of Value," in his *Radical Monotheism and Western Culture with Supplementary Essays* (New York: Harper Torchbook, 1970), p. 103.

51　Ibid., p. 106.

5 MORAL VALUES AND THE IMPERATIVE OF RESPONSIBILITY

1 John Kekes, "Constancy and Purity," in *Mind*, 92 (1983), 499.
2 J. L. Mackie, *Ethics: Inventing Right and Wrong* (Harmondsworth: Penguin Books, 1977), p. 15.
3 Richard Rorty, *Objectivity, Relativism, and Truth: Philosophical Papers*, 1 (Cambridge University Press, 1991), p. 197.
4 David O. Brink, *Moral Realism and the Foundations of Ethics* (Cambridge University Press, 1990), p. 31. Also see Sabian Lovibond, *Realism and Imagination in Ethics* (Minneapolis: University of Minnesota Press, 1983); Panayot Butchvarov, "Realism in Ethics," in *Midwest Studies in Philosophy*, XII (Minneapolis: University of Minnesota Press, 1988), pp. 395–412; Simon Blackburn, "How to Be an Ethical Antirealist," in *Midwest Studies in Philosophy*, XII (Minneapolis: University of Minnesota Press, 1988), pp. 361–376; and Geoffrey Sayre-McCord (ed.), *Essays on Moral Realism* (Ithaca: Cornell University Press, 1988).
5 Erazim Kohák, "Perceiving the Good," unpublished manuscript, p. 7. Also see his *The Embers and the Stars: An Inquiry into the Moral Sense of Nature* (University of Chicago Press, 1984).
6 Hilary Putnam, *The Many Faces of Realism* (LaSalle: Open Court, 1987), p. 17.
7 Franklin I. Gamwell, *The Divine Good: Modern Moral Theory and the Necessity of God* (San Francisco: HarperCollins, 1990), p. 101. Also see his "Moral Realism and Religion," in *The Journal of Religion*, 73:4 (1993), 475–495.
8 See Josef Fuchs, *Christian Morality: The Word Becomes Flesh*, translated by Brian McNeil (Washington: Georgetown University Press, 1981).
9 Paul Ricoeur, *Interpretation Theory: Discourse and the Surplus of Meaning* (Fort Worth: Texas Christian University, 1976).
10 See Bernard Williams "A Critique of Utilitarianism," in J. J. C. Smart and Bernard Williams (eds.), *Utilitarianism – For and Against* (Cambridge University Press, 1973), pp. 77–150.
11 Bernard Williams, "Persons, Character and Morality," in A. Rorty (ed.), *The Identities of Persons* (Berkeley: University of California Press, 1976), p. 209.
12 Charles Fried, *Right and Wrong* (Cambridge: Harvard University Press, 1978), p. 8.
13 Ibid., p. 9.
14 I am actually reworking from the perspective of an agentic-relational view of persons a long-standing position in Western ethics. Thomas Aquinas argued that there are goods humans share with other animals, other goods grounded in the social nature of human life,

and, finally, the distinctive goods of human reason. We are rational, social, animals. Plato and Augustine explored the proper ordering of the soul in terms of reason, passions, and "will," or the spirited part, to make the point. A multidimensional theory of value is coordinate to a claim about human beings as complex creatures seeking wholeness.

15 Lynne McFall, "Integrity," in John Deigh (ed.), *Ethics and Personality: Essays in Moral Psychology* (University of Chicago Press, 1992), p. 85.

16 Ibid., p. 88.

17 Germain Grisez, *The Way of the Lord Jesus, vol. 1 : Christian Moral Principles* (Chicago: Franciscan Herald Press, 1983); John Finnis, *Fundamentals of Ethics* (Washington: Georgetown University Press, 1983); and William E. May, *Moral Absolutes: Catholic Tradition, Current Trends and the Truth*, The Père Marquette Lecture in Theology 1989 (Milwaukee: Marquette University Press, 1989). For a discussion of the debate over proportionalism in Roman Catholic ethics see Bernard Hoose, *Proportionalism: The American Debate and its European Roots* (Washington: Georgetown University Press, 1987).

18 Germain Grisez and Russell Shaw, *Fulfillment in Christ* (University of Notre Dame Press, 1991), p. 80.

19 Hans Jonas, *The Imperative of Responsibility: In Search of an Ethics for the Technological Age*, translated by Hans Jonas and David Herr (University of Chicago Press, 1984), p. 85.

20 Charles Fried, *Right and Wrong*, p. 29.

21 Kristine A. Culp, "The Nature of Christian Community," unpublished chapter for a volume on feminist theology edited by Rita Nakashima Brock, Claudia Camp, and Serene Jones (St. Louis: Chalice Press, forthcoming), p. 13 of chapter manuscript.

22 See, for example, John Calvin's analysis of the decalogue in *Institutes of the Christian Religion*, 2 vols. (Philadelphia: The Westminster Press, 1960).

23 Jeffrey Blustein, *Care and Commitment: Taking the Personal Point of View* (Oxford University Press, 1991), p. 116.

6 FREEDOM AND RESPONSIBILITY

1 Peter A. French, *Responsibility Matters* (Lawrence: University Press of Kansas, 1992), p. 3.

2 J. R. Lucas, *Responsibility* (Oxford: Clarendon Press, 1993), p. 5.

3 Wolfgang Huber, *Konflikt und Konsens: Studien zur Ethik der Verantwortung* (München: Chr. Kaiser, 1990).

4 Isaiah Berlin, *Two Concepts of Liberty* (Oxford University Press, 1958), pp. 6–7.

5 In order to affirm the principle of (negative) freedom, any decision to constrict freedom or private right must rest on the free consent of all parties in the social contract. See John Rawls, *The Theory of Justice* (Cambridge: Harvard University Press, 1971).

6 See Timothy Jackson, "Liberalism and *Agape*: The Priority of Charity to Democracy and Liberalism," in *Annual of the Society of Christian Ethics* (Washington: Georgetown University Press, 1993), pp. 47–72.

7 Berlin, *Two Concepts of Liberty*, p. 7.

8 John Stuart Mill, *An Examination of Sir William Hamilton's Philosophy*, 3rd ed. (London: Longmans, Green, Reader, and Dyer, 1867).

9 Berlin, *Two Concepts of Liberty*, p. 16.

10 Charles Taylor, *Philosophical Papers, vol. 2: Philosophy and the Human Sciences* (Cambridge University Press, 1985), p. 213.

11 Ibid., p. 216.

12 Lucas, *Responsibility*, p. 86. For a discussion of theories of punishment also see H. B. Acton (ed.), *The Philosophy of Punishment* (London: Macmillan, 1969).

13 See the introduction by John Martin Fischer in *Moral Responsibility*, edited by John Martin Fischer (Ithaca: Cornell University Press, 1986), pp. 9–64.

14 One can speak of both causal and semantic explanation. Causal explanation is concerned with the determining cause and consequences of actions, while semantic explanation centers on the meaning of terms and the logic of sign systems. In terms of theories of freedom, the problem of causal explanation has played the greater role in modern ethics.

15 Gary Watson, "Free Agency," in *Moral Responsibility*, p. 85.

16 Kenneth E. Kirk, *The Threshold of Ethics* (London: Skeffington and Son, Ltd., n.d.), pp. 61–62.

17 Albert R. Jonsen, *Responsibility in Modern Religious Ethics* (Washington: Corpus Books, 1968), pp. 35–74.

18 Ibid., p. 38.

19 Lucas, *Responsibility*, p. 9.

20 Martin Luther, "Secular Authority: To What Extent It Should Be Obeyed," in John Dillenberger, *Martin Luther: Selections from His Writings* (Garden City: Anchor Books, 1961), pp. 363–402.

21 Jonsen, *Responsibility in Modern Religious Ethics*, p. 42. My discussion has been greatly aided by Jonsen's analysis, even while I develop the argument in a distinctive way.

22 Aristotle, *Nichomachean Ethics*, Book III. 3. 1112b20.

23 See John Howard Yoder, *The Politics of Jesus: Vicit Agnus Noster* (Grand Rapids: William B. Eerdmans, 1972) and also Stanley

Hauerwas, *Christian Existence Today: Essays on Church, World, and Living In Between* (Durham: Labyrinth Press, 1988).

24 Jonsen, *Responsibility in Modern Religious Ethics*, p. 69.

25 Lucas, *Responsibility*, p. 11.

7 RESPONSIBILITY AND MORAL IDENTITY

1 Thomas Aquinas, *Summa Theologiae*, 1/11 q. 58, art. 1.

2 Ibid., q. 58, art. 1, obj. 2.

3 The extreme case here is, of course, warfare. The state can enlist the lives of citizens for purposes of defense. Offensive wars are categorically immoral even though there may be justifiable cause for a defensive war. See Richard B. Miller, *Interpretations of Conflicts: Ethics, Pacifism, and the Just War Tradition* (University of Chicago Press, 1991).

4 See Thomas Aquinas, *Summa Theologiae*, 1/11, q. 62, art. 2.

5 Peter A. French, *Responsibility Matters* (Lawrence: University Press of Kansas, 1992), p. 65.

6 Ibid., p. 67.

7 See Amitai Etzioni, *The Moral Dimension: Toward a New Economics* (New York: The Free Press, 1988).

8 Iris Murdoch, *Metaphysics as a Guide to Morals* (London: The Penguin Press, Allen Lane, 1992), p. 188.

9 See Paul Ricoeur, *Interpretation Theory: Discourse and the Surplus of Meaning* (Fort Worth: Texas Christian Press, 1976).

10 Susan Wolf, "Sanity and the Metaphysics of Responsibility," in John Christman (ed.), *The Inner Citadel: Essays on Individual Autonomy*, (New York: Oxford University Press, 1989), p. 140.

11 Roman Ingarden, *Man and Value*, translated by Arthur Szylewicz (Washington: The Catholic University Press of America, 1983), p. 78.

12 See Hans-Georg Gadamer, *Truth and Method*, revised translation by Joel B. Weinsheimer and Donald G. Marshall (New York: Continuum, 1989).

13 See Alasdair MacIntyre, *Three Rival Versions of Moral Enquiry: Encyclopedia, Genealogy, and Tradition* (University of Notre Dame Press, 1990). Also see Jean Porter, "Openness and Constraint: Moral Reflection as Tradition-Guided Inquiry in Alasdair MacIntyre's Recent Works," in *The Journal of Religion*, 73:4 (1993), 514–536.

14 See Charles Taylor, *Sources of the Self: The Making of Modern Identity* (Cambridge: Harvard University Press, 1989) and his "Responsibility for Self," in Gary Watson (ed.), *Free Will* (Oxford University Press, 1982), pp. 111–126.

15 Harry Frankfurt calls individuals who do not make these judgments

"wantons." These are "agents who have first-order desires but who are not persons because, whether or not they have desires of the second order, they have no second-order volitions." See Harry Frankfurt, *The Importance of What We Care About: Philosophical Essays* (Cambridge University Press, 1988), p. 16.

16 Charles Taylor, "Responsibility for Self," p. 123.
17 Ibid., p. 125.
18 Taylor, *Sources of the Self*, p. 517.
19 Ibid., p. 449.
20 Ibid., p. 448.
21 Alan Donagan, *The Theory of Morality* (The University of Chicago Press, 1977), p. 242.
22 See William Schweiker, "Accounting for Ourselves: Accounting Practice and the Discourse of Ethics," in *Accounting, Organizations and Society*, 18: 2/3 (1993), 231–252.
23 See Robert P. Scharlemann, *The Reason of Following: Christology and the Ecstatic I* (University of Chicago Press, 1991).
24 My argument here is analogous to that of Karl Rahner on the presence of the symbol "God" in Western thought. Rahner argues that the presence of this symbol in our culture signals the openness of human life to a horizon of absolute mystery. I am arguing that the presence of the name of God in our culture grounds a belief that the exercise of power itself is not the human good. See Karl Rahner, *Grace in Freedom* (New York: Herder and Herder, 1969), pp. 183–203.
25 See Friedrich Nietzsche, *The Birth of Tragedy and the Genealogy of Morals*, translated by Francis Golffing (New York: Doubleday Anchor Books, 1956).
26 On collective responsibility, see essays by D. E. Cooper, Virginia Held, and Michael J. Zimmerman in Peter A. French (ed.), *The Spectrum of Responsibility* (New York: St. Martin's Press, 1991), pp. 251–286.
27 See Milton Friedman, "The Social Responsibility of Business Is to Increase Its Profits," in *New York Times Magazine*, 13 (September 1970), 32 ff.; Thomas Mulligan, "A Critique of Milton Friedman's Essay 'The Social Responsibility of Business Is to Increase Its Profits,'" in *Journal of Business Ethics*, 5 (1986), 265–269; and Paul Weaver, "After Social Responsibility," in John R. Meyer and James M. Gustafson (eds.), *The U.S. Business Corporation: An Institution in Transition* (Cambridge: Ballinger Publishing, 1988), pp. 133–148.
28 Amartya K. Sen, "Rational Fools: A Critique of the Behavioral Foundations of Economic Theory," in Frank Hahn and Martin Hollis (eds.), *Philosophy and Economic Theory* (Oxford University Press, 1979), p. 87.

29 Reinhold Niebuhr, *Moral Man and Immoral Society* (New York: Charles Scribner's Sons, 1932).

30 Peter A. French, "Corporate Moral Agency," in Michael Hoffman and Jennifer Mills Moore (eds.), *Business Ethics: Readings and Cases in Corporate Morality* (New York: McGraw-Hill, 1984), pp. 163–171. Also see his "The Corporation as a Moral Person," in Peter French (ed.), *The Spectrum of Responsibility* (New York: St. Martin's Press, 1991), pp. 290–304. For responses to French's argument in the same volume see John Ladd, "Corporativism," pp. 305–312, and Larry May, "Vicarious Agency and Corporate Responsibility," pp. 313–324.

31 John R. Danley, "Corporate Moral Agency: The Case for Anthropological Bigotry," in Michael Hoffman and Jennifer Mills Moore (eds.), *Business Ethics: Readings and Cases in Corporate Morality* (New York: McGraw-Hill, 1984), pp. 172–179.

8 POWER, RESPONSIBILITY, AND THE DIVINE

1 Christian theologians earlier in this century developed dialogical theories of responsibility by drawing on the thought of the Jewish philosopher Martin Buber. By addressing the ethics of Hans Jonas, a post-holocaust philosopher, I hope not only to reconstruct the ethics of responsibility, but also to insist that Christian theology continually affirm its relation with Jewish faith and life.

2 Moses Mendelssohn, "Über die Frage: was heisst aufklären?" in *Was ist Aufklaren: Kant, Erhard, Hamann, Herder, Lessing, Mendelssohn, Reim, Schiller, Wieland*, herausgegeben von Ehrhard Bahr (Stuttgart: Philipp Reclam, 1974), p. 4. Also see Jürgen Habermas, *The Philosophical Discourse of Modernity: Twelve Lectures*, translated by Frederick Lawrence (Boston: MIT Press, 1987) and Johann Baptist Metz, Jürgen Moltmann, and Willi Oelmüller, *Kirche im Prozess der Aufklärung: Aspekte einer neuen "politischen Theologie"* (München: Chr. Kaiser, 1970).

3 Paul Ricoeur, *Oneself as Another*, translated by Kathleen Blamey (University of Chicago Press, 1992), p. 336. Also see William Schweiker, "Imagination, Violence, and Hope: A Theological Response to Ricoeur's Moral Philosophy," in David E. Klemm and William Schweiker (eds.), *Meanings in Texts and Actions: Questioning Paul Ricoeur* (Charlottesville: University Press of Virginia, 1993), pp. 205–225.

4 On this problem see Hans-Georg Gadamer, "Gibt es auf Erden ein Mass?" in *Philosophische Rundschau*, 31: 3/4 (1984), 161–177 and "Gibt es auf Erden ein Mass? (Fortsetzung)" in *Philosophische Rundschau*, 32: 1/2 (1985), 1–25.

5 On this see G. E. M. Anscombe, "Modern Moral Philosophy," in *Philosophy*, 33 (1958), 1–19 and Kai Nielsen, *Ethics without God*, rev. edn. (Buffalo: Prometheus Books, 1990).

6 Hans Jonas, *Technik, Medizin und Ethik: zur Praxis des Prinzips Verantwortung* (Frankfurt: Suhrkamp, 1987), p. 302.

7 Ibid.

8 See Hans Jonas, "The Concept of God After Auschwitz: A Jewish Voice," in *The Journal of Religion*, 67:1 (1987), 1–13.

9 Hans Jonas, *The Imperative of Responsibility: In Search of an Ethic for the Technological Age*, translated by Hans Jonas and David Herr (University of Chicago Press, 1984), p. 81. For Jonas' metaphysics see his *The Phenomenon of Life: Towards a Metaphysical Biology* (Chicago: Midway, 1982). Also see Strachan Donnelley, "Whitehead and Hans Jonas: Organism, Causality, and Perception," in *International Philosophical Quarterly*, 19:3 (1979), 301–315 and T. A. Goudge, "Existentialism and Biology," in *Dialogue: Canadian Philosophical Review*, 5:4 (1967), 603–608. For a critical response to Jonas' ethics from the perspective of discourse ethics see Karl-Otto Apel, *Diskurs und Verantwortung: Das Problem des übergangs zur postkonventionellen Moral* (Frankfurt: Suhrkamp, 1990), pp. 179–218.

10 Jonas, *The Imperative of Responsibility*, p. 11.

11 Ibid., p. 43. See also James M. Gustafson, "Theology and Ethics: An Interpretation of the Agenda," and Hans Jonas, "Response to James M. Gustafson," in H. Tristain Englehardt, Jr. and Daniel Callahan (eds.), *Knowing and Valuing: The Search for Common Roots* (New York: The Hasting Center: Institute for Society, Ethics, and the Life Sciences, 1980), pp. 181–217.

12 Ibid., p. 79.

13 Ibid., pp. 89–90.

14 Ibid., p. 85.

15 Ibid., p. 130.

16 Ibid., p. 87.

17 Ibid., p. 12.

18 Ibid., p. 85.

19 See Beverly Harrison, *Making the Connections: Essays in Feminist Social Ethics* (Boston: Beacon Press, 1985), pp. 235–266.

20 See Hans Jonas, "The Concept of God After Auschwitz."

21 Jonas' argument should be seen in contrast to claims about the absent, but coming, eschatological God found in Nietzsche and also in Heidegger's claims about the fate of being. This account draws on the image of Dionysus, the god who is coming, and also Christ while portraying our age as the "night of the gods." "Nietzsche had entrusted the overcoming of nihilism to the aesthetically revived

Dionysian myth. Heidegger projects this Dionysian happening onto the scene of a critique of metaphysics, which thereby takes on world-historical significance ... As a result, Being can only come about as a fateful dispensation; those who are in need can at most hold themselves open and prepare for it." Habermas, *The Philosophical Discourse of Modernity*, p. 99.

22 On the ontological proof see Paul Tillich, *Systematic Theology*, 3 volumes in 1 (University of Chicago Press, 1967) and Iris Murdoch, *Metaphysics as a Guide to Morals* (London: The Penguin Press, Allen Lane, 1993).

23 The "community of life" obviously includes non-living entities on which life depends, for example, water, light, oxygen. The imperative that we are to respect and enhance the integrity of life at the planetary level encompasses the complex interpendent reality of the ecosphere.

9 RESPONSIBILITY AND CHRISTIAN ETHICS

1 On the theological ethics of culture, see William Schweiker, "Hermeneutics, Ethics, and the Theology of Culture: Concluding Reflections" in David E. Klemm and William Schweiker (eds.), *Meanings in Texts and Actions: Questioning Paul Ricoeur* (Charlottesville: The University Press of Virginia, 1993), pp. 292–313.

2 H. Richard Niebuhr, *The Responsible Self: An Essay in Christian Moral Philosophy* (New York: Harper and Row, 1963), pp. 42–44.

3 See St. Augustine, *The City of God*, XIX, 14 (New York: Doubleday Image Books, 1958).

4 On this see Jürgen Habermas, *Moral Consciousness and Communicative Action*, translated by Christian Lenhardt and Shierry Weber Nicholsen with an introduction by Thomas McCarthy (Cambridge: The MIT Press, 1990). Also see Karl-Otto Apel, "Discourse Ethics as a Response to the Novel Challenges of Today's Reality to Coresponsibility," in *The Journal of Religion*, 73:4 (1993), 496–513.

5 See Alasdair MacIntyre, *Three Rival Versions of Moral Enquiry: Encyclopedia, Genealogy and Tradition* (University of Notre Dame Press, 1990).

6 Charles Taylor, *Sources of the Self: The Making of Modern Identity* (Cambridge: Harvard University Press, 1989), p. 505.

7 Thomas Aquinas, *Summa Theologiae*, Ia, q. 44, art. 1.

8 Reinhold Niebuhr, *An Interpretation of Christian Ethics* (New York: Seabury Press, 1979), p. 3.

Select bibliography

Acton, H. B. (ed.), *The Philosophy of Punishment* (London: Macmillan, 1969).
Adkins, A. W. H., *Merit and Responsibility* (Oxford University Press, 1960).
André Lelande (ed.), *Vocabulaire technique et critique de la Philosophie* 6th edn., (Paris, 1951).
Anscombe, Elizabeth, "Modern Moral Philosophy," in *Philosophy* 33 (1958).
Apel, Karl-Otto, "Discourse Ethics as a Response to the Novel Challenges of Today's Reality to Co-responsibility," in *The Journal of Religion* 73:4 (1993).
  Diskurs und Verantwortung: Das Problem des Übergangs zur postkonventionellen Moral (Frankfurt: Suhrkamp, 1990).
Aquinas, Thomas, *Summa Theologiae* 5 vols., translated by The English Dominican Province (Westminster: Christian Classics, 1981).
Aristotle, *Nicomachean Ethics* in Richard McKeon (ed.), *Introduction to Aristotle* (New York: The Modern Library, 1947), Book II.
Augustine, St., *The City of God* (New York: Doubleday Image Books, 1958).
Barth, Karl, *Church Dogmatics* edited by G. W. Bromiley and T. F. Torrance (Edinburgh: T. and T. Clark, 1957-1970).
Berlin, Isaiah, *Two Concepts of Liberty* (Oxford University Press, 1958).
Blustein, Jeffrey, *Care and Commitment: Taking the Personal Point of View* (Oxford University Press, 1991).
Bonhoeffer, Dietrich, *Ethics* edited by Eberhard Bethge (New York: Collier Books, 1986).
Bradley, F. H., *Ethical Studies* (Oxford University Press, 1988).
Brink, David O., *Moral Realism and the Foundations of Ethics* (Cambridge University Press, 1990).
Brunner, Emil, *Truth as Encounter* (Philadelphia: The Westminster Press, 1964).
Buber, Martin, *I and Thou* (Edinburgh: T. and T. Clark, 1937).

Calvin, John, *Institutes of the Christian Religion*, 2 vols., edited by John T. McNeill (Philadelphia: The Westminster Press, 1960).

Curran, Charles, *Directions in Fundamental Moral Theology* (University of Notre Dame Press, 1985).

Donagan, Alan, *The Theory of Morality* (University of Chicago Press, 1977).

Emmet, Dorothy, *Rules, Roles and Relations* (New York: St. Martin's Press, 1967).

The Moral Prism (London: Macmillan Press, 1979).

Etzioni, Amitai, *The Moral Dimension: Toward a New Economics* (New York: The Free Press, 1988).

Finnis, John, *Fundamentals of Ethics* (Washington: Georgetown University Press, 1983).

Fischer, John Martin (ed.), *Moral Responsibility* introduction by John Martin Fischer (Ithaca: Cornell University Press, 1986).

Frankena, William K., *Thinking about Morality* (Ann Arbor: The University of Michigan Press, 1980).

Frankfurt, Harry, *The Importance of What We Care About: Philosophical Essays* (Cambridge University Press, 1988).

Freire, Paulo, *The Pedagogy of the Oppressed* (New York: Continuum, 1970).

French, Peter A., *Responsibility Matters* (Lawrence: University Press of Kansas, 1992).

Fried, Charles, *Right and Wrong* (Cambridge: Harvard University Press, 1978).

Fuchs, Josef, *Christian Morality: The Word Becomes Flesh* translated by Brian McNeil (Washington: Georgetown University Press, 1981).

Gadamer, Hans-Georg, *Truth and Method* revised translation by Joel B. Weinsheimer and Donald G. Marshall (New York: Continuum, 1989).

Gamwell, Franklin I., *The Divine Good: Modern Moral Theory and the Necessity of God* (San Francisco: HarperCollins, 1990).

Gilkey, Langdon, *Society and the Sacred: Towards a Theology for a Culture in Decline* (New York: Crossroads, 1981).

Girard, René, *Violence and the Sacred* translated by Patrick Gregory (Baltimore: Johns Hopkins University Press, 1977).

Grisez, Germain, *The Way of the Lord Jesus vol. 1 : Christian Moral Principles* (Chicago, 1983).

Grisez, Germain, and Shaw, Russell, *Fulfillment in Christ* (University of Notre Dame Press, 1991).

Gula, Richard, *What Are They Saying About Moral Norms?* (New York: Paulist Press, 1982).

Gustafson, James M., *A Sense of the Divine: The Natural Environment from a Theocentric Perspective* (Cleveland: The Pilgrim Press, 1994).

Ethics from a Theocentric Perspective 2 vols. (University of Chicago Press, 1981, 1984).

Protestant and Roman Catholic Ethics: Prospects for Rapprochement (University of Chicago Press, 1978).

Gutman, James, "Integrity as a Standard of Evaluation," in *Journal of Philosophy* 42 (1945).

Habermas, Jürgen, *Moral Consciousness and Communicative Action* translated by Christian Lenhardt and Shierry Weber Nicholsen with an introduction by Thomas McCarthy (Cambridge: The MIT Press, 1990).

The Philosophical Discourse of Modernity: Twelve Lectures translated by Frederick Lawrence (Cambridge: MIT Press, 1987).

Halfon, Mark S., *Integrity: A Philosophical Analysis* (Philadelphia: Temple University Press, 1989).

Harrison, Beverly Wildung, *Making the Connections: Essays in Feminist Social Ethics* (Boston: Beacon Press, 1985).

Hauerwas, Stanley, *Vision and Virtue: Essays in Christian Ethical Reflection* (University of Notre Dame Press, 1974).

A Community of Character: Toward a Constructive Christian Social Ethic (University of Notre Dame Press, 1981).

Against the Nations: War and Survival in a Liberal Society (Minneapolis: Winston, 1985).

Christian Existence Today: Essays on Church, World, and Living In Between (Durham: Labyrinth Press, 1988).

Heidegger, Martin, *Being and Time* translated by John Macquarrie and Edward Robinson (New York: Harper and Row, 1962).

Henriot, Jacques, "Responsabilité" in Sylvain Auroux (ed.), *Encyclopédie Philosophique Universelle II Les Notions Philosophiques Doctinionaire* (Paris: Presses Universitaires de France, vol. II, 1989).

Hoagland, Sarah Lucia, *Lesbian Ethics: Toward New Values* (Palo Alto: Institute of Lesbian Studies, 1988).

Hoose, Bernard, *Proportionalism: The American Debate and its European Roots* (Washington: Georgetown University Press, 1987).

Huber, Wolfgang, "Towards an Ethics of Responsibility," in *The Journal of Religion* 73:4 (1993).

Konflikt und Konsens: Studien zur Ethik der Verantwortung (München: Chr. Kaiser, 1990).

Ingarden, Roman, *Über die Verantwortung: Ihre ontischen Fundamente* (Stuttgart: Reclam, 1970).

Jonas, Hans, "Contemporary Problems in Ethics from a Jewish Perspective," in his *Philosophical Essays: From Ancient Creed to Technological Man* (Englewoood Cliffs: Prentice-Hall, 1974).

"The Concept of God After Auschwitz: A Jewish Voice" in *The Journal of Religion* 67:1 (1987).

The Phenomenon of Life: Towards a Metaphysical Biology (Chicago: Midway, 1982).

The Imperative of Responsibility: In Search for an Ethics of the Technological Age translated by Hans Jonas and David Herr, (University of Chicago Press, 1984).

Technik, Medizin und Ethik: zur Praxis des Prinzips Verantwortung (Frankfurt: Suhrkamp, 1987).

Jonsen, Albert R., *Responsibility in Modern Religious Ethics* (Washington: Corpus Books, 1968).

Kant, Immanuel, *Fundamental Principles of the Metaphysics of Morals* translated by Thomas K. Abbot (New York: Liberal Arts Press, 1949).

Kekes, John, "Constancy and Purity," in *Mind* 92 (1983).

Kirk, Kenneth E., *The Threshold of Ethics* (London: Skeffington and Son, Ltd., n.d.).

Kohák, Erazim, *The Embers and the Stars: An Inquiry into the Moral Sense of Nature* (University of Chicago Press, 1984).

Küng, Hans, *Global Responsibility: In Search of a New Global Ethic* translated by John Bowden (New York: Crossroads, 1991).

Lange, Dietz, *Ethik in evangelischer Perspektive* (Göttingen: Vandenhoeck and Ruprecht, 1992).

Levinas, Emmanuel, *Totality and Infinity: An Essay on Exteriority* translated by A. Lingis (Pittsburgh: Duquesne University, 1969).

Lévy-Bruhl, L., *L'Idée de Responsibilité* (Paris: Librairie Hachette et Cie, 1844).

Lovibond, Sabian, *Realism and Imagination in Ethics* (Minneapolis: University of Minnesota Press, 1983).

Lucas, J. R., *Responsibility* (Oxford: Clarendon Press, 1993).

MacIntyre, Alasdair, *After Virtue: A Study in Moral Theory* (University of Notre Dame Press, 1981).

Three Rival Versions of Moral Enquiry: Encyclopedia, Genealogy, and Tradition (University of Notre Dame Press, 1990).

Mackie, J. L., *Ethics: Inventing Right and Wrong* (Harmondsworth: Penguin Books, 1977).

May, William E., *Moral Absolutes: Catholic Tradition, Current Trends and the Truth* The Père Marquette Lecture in Theology 1989 (Milwaukee: Marquette University Press, 1989).

McFall, Lynne, "Integrity," in John Deigh (ed.) *Ethics and Personality: Essays in Moral Psychology* (The University of Chicago Press, 1992).

McKeon, Richard, "The Development and Significance of the Concept of Responsibility," in his *Freedom and History and Other Essays* (University of Chicago Press, 1990).

Midgley, Mary, *Can't We Make Moral Judgments?* (New York: St. Martin's Press, 1992).

Select bibliography

Mill, John Stuart, *An Examination of Sir William Hamilton's Philosophy*, 3rd
edn. (London: Longmans, Green, Reader, and Dyer, 1867).

Mount, Eric Jr., *Conscience and Responsibility* (Richmond: John Knox
Press, 1969).

Murdoch, Iris, *Metaphysics as a Guide to Morals* (London: The Penguin
Press, Allen Lane, 1992).

Niebuhr, H. Richard, *The Responsible Self: An Essay in Christian Moral
Philosophy*, introduction by James M. Gustafson (New York: Harper
and Row, 1963).

Radical Monotheism and Western Culture with Supplementary Essays (New
York: Harper Torchbook, 1970).

Niebuhr, Reinhold, *Moral Man and Immoral Society* (New York: Charles
Scribner's Sons, 1932).

An Interpretation of Christian Ethics (New York: Seabury Press, 1979).

Nielsen, Kai, *Ethics without God* rev. edn. (Buffalo: Prometheus Books,
1990).

Nietzsche, Friedrich, *The Birth of Tragedy and the Genealogy of Morals*
translated by Francis Golffing (New York: Doubleday Anchor
Books, 1956).

Picht, Georg, "Der Begriff der Verantwortung," in his *Wahrheit, Vernunft,
Verantwortung: philosophische Studien* (Stuttgart: Ernst Klett, 1969).

Putnam, Hilary, *The Many Faces of Realism* (LaSalle: Open Court, 1987).

Rahner, Karl, *Grace in Freedom* (New York: Herder and Herder, 1969).

Ramsey, Paul, *Basic Christian Ethics* (New York: Charles Scribner's Sons
1950).

Fabricated Man: The Ethics of Genetic Control (New Haven: Yale
University Press, 1970).

Rawls, John, *A Theory of Justice* (Cambridge: Harvard University Press,
1971).

Ricoeur, Paul, "Naming God," in *Union Seminary Quarterly Review* 34
(1979).

"The Golden Rule: Exegetical and Theological Perplexities," in *New
Testament Studies* 36 (1990).

Interpretation Theory: Discourse and the Surplus of Meaning (Fort Worth:
Texas Christian University Press, 1976).

Soi-même comme un autre (Paris: Éditions du Seuil, 1990).

Rorty, Richard, *Contingency, Irony, and Solidarity* (Cambridge University
Press, 1989).

Sartre, Jean-Paul, *Being and Nothingness: An Essay on Phenomenological
Ontology* translated by Hazel Barnes (New York: Philosophical
Library, 1956).

Sayre-McCord, Geoffrey (ed.), *Essays on Moral Realism* (Ithaca: Cornell
University Press, 1988).

Scharlemann, Robert P., *The Reason of Following: Christology and the Ecstatic I* (University of Chicago Press, 1991).

Schweiker, William, "The Good and Moral Identity: A Theological Ethical Response to Charles Taylor's, *Sources of the Self*," in *The Journal of Religion*, 72:4 (1992).

"Accounting for Ourselves: Accounting Practice and the Discourse of Ethics," in *Accounting, Organizations and Society* 18:2/3 (1993).

"Hermeneutics, Ethics, and the Theology of Culture: Concluding Reflections," in David E. Klemm and William Schweiker (eds.), *Meanings in Texts and Actions: Questioning Paul Ricoeur* (Charlottesville: The University Press of Virginia, 1993).

"Imagination, Violence, and Hope: A Theological Response to Ricoeur's Moral Philosophy," in David E. Klemm and William Schweiker (eds.), *Meanings in Texts and Actions: Questioning Paul Ricoeur* (Charlottesville: University Press of Virginia, 1993).

"Radical Interpretation and Moral Responsibility: A Proposal for Theological Ethics," in *The Journal of Religion* 73:4 (1993).

Mimetic Reflections: A Study in Hermeneutics, Theology, and Ethics (New York: Fordham University Press, 1990).

Schweiker, William and Per M. Anderson (eds.), *Worldviews and Warrants: Plurality and Authority in Theology* (Lanham: University Press of America, 1987).

Sen, Amartya K., "Rational Fools: A Critique of the Behavioral Foundations of Economic Theory," in Frank Hahn and Martin Hollis (eds.), *Philosophy and Economic Theory* (Oxford University Press, 1979).

Shklar, Judith, *Legalism* (Cambridge: Harvard University Press, 1964).

Smiley, Marion, *Moral Responsibility and the Boundaries of Community: Power and Accountability from a Pragmatic Point of View* (University of Chicago Press, 1992).

Stout, Jeffrey, *Ethics after Babel: The Languages of Morals and their Discontents* (Boston: Beacon Press, 1988).

Taylor, Charles, "Responsibility for Self," in Gary Watson (ed.), *Free Will* (Oxford University Press, 1982).

Philosophical Papers, 2 vols. (Cambridge University Press, 1985).

Sources of the Self: The Making of Modern Identity (Cambridge: Harvard University Press, 1990).

The Ethics of Authenticity (Cambridge: Harvard University Press, 1992).

Taylor, Gabriele, "Integrity," in *Proceedings of the Aristotelian Society* 55 (1981).

Tillich, Paul, *Morality and Beyond* (New York: Harper Torchbook, 1963).

Systematic Theology, 3 volumes in 1 (University of Chicago Press, 1967).

Tracy, David, *Plurality and Ambiguity: Hermeneutics, Religion, Hope* (San Francisco: Harper and Row, 1986).

Troeltsch, Ernst, *The Social Teachings of the Christian Churches*, 2 vols., in The Library of Theological Ethics (Louisville: John Knox/Westminster Press, 1992).

Wartenberg, Thomas F., *The Forms of Power: From Domination to Transformation* (Philadelphia: Temple University Press, 1990).

Watson, Gary, "Free Agency," in John Martin Fischer (ed.), *Moral Responsibility* introduction by John Martin Fischer (Ithaca: Cornell University Press, 1986).

Weber, Max, "Politik als Beruf," in *Gesammelte politische Schriften*, herausgegeben von Johannes Winckelmann, 4th edn. (Tübingen: J. C. B. Mohr [Paul Siebeck], 1980).

Weischedel, Wilhelm, *Das Wesen der Verantwortung: Ein Versuch* Dritte unveränderte Auflage (Frankfurt: Vittorio Klostermann, 1972).

Williams, Bernard, "Persons, Character and Morality," in A. Rorty, (ed.), *The Identities of Persons* (Berkeley: University of California Press, 1976).

Ethics and the Limits of Philosophy (Cambridge: Harvard University Press, 1985).

Wolf, Susan, "Sanity and the Metaphysics of Responsibility," in John Christman (ed.), *The Inner Citadel: Essays on Individual Autonomy* (New York: Oxford University Press, 1989).

Yoder, John Howard, *The Politics of Jesus: Vicit Agnus Noster* (Grand Rapids: William B. Eerdmans, 1972).

Index

NEW STUDIES IN CHRISTIAN ETHICS

Printed in the USA
CPSIA information can be obtained
at www.ICGtesting.com
LVHW092043191223
766947LV00008B/81